RADIO, SUBMILLIMETER, AND X-RAY TELESCOPES

RADIOTELESKOPY. SUBMILLIMETROVYE I RENTGENOVSKIE TELESKOPY

РАДИОТЕЛЕСКОПЫ. СУБМИЛЛИМЕТРОВЫЕ И РЕНТГЕНОВСКИЕ ТЕЛЕСКОПЫ

The Lebedev Physics Institute Series

Editors: Academicians D. V. Skobel'tsyn and N. G. Basov

P. N. Lebedev Physics Institute, Academy of Sciences of the USSR

Recent Volumes in this Series

Volume 35	Electronic and Vibrational Spectra of Molecules
Volume 36	Photodisintegration of Nuclei in the Giant Resonance Region
Volume 37	Electrical and Optical Properties of Semiconductors
Volume 38	Wideband Cruciform Radio Telescope Research
Volume 39	Optical Studies in Liquids and Solids
Volume 40	Experimental Physics: Methods and Apparatus
Volume 41	The Nucleon Compton Effect at Low and Medium Energies
Volume 42	Electronics in Experimental Physics
Volume 43	Nonlinear Optics
Volume 44	Nuclear Physics and Interaction of Particles with Matter
Volume 45	Programming and Computer Techniques in Experimental Physics
Volume 46	Cosmic Rays and Nuclear Interactions at High Energies
Volume 47	Radio Astronomy: Instruments and Observations
Volume 48	Surface Properties of Semiconductors and Dynamics of Ionic Crystals
Volume 49	Quantum Electronics and Paramagnetic Resonance
Volume 50	Electroluminescence
Volume 51	Physics of Atomic Collisions
Volume 52	Quantum Electronics in Lasers and Masers, Part 2
Volume 53	Studies in Nuclear Physics
Volume 54	Photomesic and Photonuclear Reactions and Investigation Method with Synchrotrons
Volume 55	Optical Properties of Metals and Intermolecular Interactions
Volume 56	Physical Processes in Lasers
Volume 57	Theory of Interaction of Elementary Particles at High Energies
Volume 58	Investigations in Nonlinear Optics and Hyperacoustics
Volume 59	Luminescence and Nonlinear Optics
Volume 60	Spectroscopy of Laser Crystals with Ionic Structure
Volume 61	Theory of Plasmas
Volume 62	Methods in Stellar Atmosphere and Interplanetary Plasma Research
Volume 63	Nuclear Reactions and Interaction of Neutrons and Matter
Volume 64	Primary Cosmic Radiation
Volume 65	Stellarators
Volume 66	Theory of Collective Particle Acceleration and Relativistic Electron Beam Emission
Volume 67	Physical Investigations in Strong Magnetic Fields
Volume 68	Radiative Recombination in Semiconducting Crystals
Volume 69	Nuclear Reactions and Charged-Particle Accelerators
Volume 70	Group-Theoretical Methods in Physics
Volume 71	Photonuclear and Photomesic Processes
Volume 72	Physical Acoustics and Optics: Molecular Scattering of Light; Propagation of Hypersound; Metal Optics
Volume 73	Microwave–Plasma Interactions
Volume 74	Neutral Current Sheets in Plasmas
Volume 75	Optical Properties of Semiconductors
Volume 76	Lasers and Their Applications
Volume 77	Radio, Submillimeter, and X-Ray Telescopes
Volume 78	Research in Molecular Laser Plasmas
Volume 79	Luminescence Centers in Crystals

In preparation

Volume 80	Synchrotron Radiation
Volume 81	Pulse Gas-Discharge Atomic and Molecular Lasers

Proceedings (Trudy) of the P. N. Lebedev Physics Institute

Volume 77

RADIO, SUBMILLIMETER, AND X-RAY TELESCOPES

Edited by
N. G. Basov

P. N. Lebedev Physics Institute
Academy of Sciences of the USSR
Moscow, USSR

Translated from Russian by
Edward U. Oldham

Springer Science+Business Media, LLC

Library of Congress Cataloging in Publication Data

Main entry under title:

Radio, submillimeter, and x-ray telescopes.

(Proceedings (Trudy) of the P. N. Lebedev Physics Institute; v. 77)
Translation of Radioteleskopy submillimetrovye i rentgenovskie telescopy.
Includes bibliographical references and index.
1. Radio telescope—Addresses, essays, lectures. 2. Infra-red telescope—Addresses,
essays, lectures. 3. X-ray telescope—Addresses, essays, lectures. I. Basov, Nikolaĭ
Gennadievich, 1922-   II. Series: Akademiĭa nauka SSSR. Fizicheskiĭ institut.
Proceedings: v. 77
QC1.A4114 vol. 77 [QB479.2] 530'.08s [522'.6]
ISBN 978-1-4899-2662-3   76-48290

ISBN 978-1-4899-2662-3   ISBN 978-1-4899-2660-9 (eBook)
DOI 10.1007/978-1-4899-2660-9

The original Russian text was published by Nauka Press in Moscow in 1974 for the
Academy of Sciences of the USSR as Volume 77 of the Proceedings of the P. N. Lebedev
Physics Institute. This translation is published under an agreement with the Copyright
Agency of the USSR (VAAP).

PREFACE

This volume contains the results of research and development connected with the creation of telescopes intended for the new regions of the spectrum mastered by astronomy: the x-ray, submillimeter (far infrared), and radio regions.

The creation of x-ray, submillimeter, and radio telescopes and the receiver apparatus connected with them is a complicated and, in many respects, unusual problem. Therefore, the experience accumulated at the Institute of Physics can prove useful to specialists working in this field.

This volume is intended for scientists, engineers, and builders occupied in research and development in the fields of x-ray, submillimeter, and radio astronomy as well as for students of advanced courses in these specialties.

CONTENTS

A Reflecting X-Ray Telescope for an Orbital Astrophysical Station.   1
  I. L. Beigman, L. A. Vainshtein, Yu. P. Voinov, D. A. Goganov,
  N. I. Komyak, S. L. Mandel'shtam, I. P. Tindo, N. A. Shatskii,
  and A. I. Shurygin

Mirror Systems for X-Ray Telescopes.   13
  I. L. Beigman, L. A. Vainshtein, Yu. P. Voinov, and V. P. Shevel'ko

Extra-Atmospheric Studies in the Submillimeter Range Using On-Board
  Telescopes   35
  A. E. Salomonovich and A. S. Khaikin

Optical Systems of On-Board Submillimeter Telescopes   59
  A. S. Khaikin

A Two-Channel Cooled Receiver for On-Board Telescopes of the Submillimeter
  Range.   85
  A. A. Kobzev, V. I. Lapshin, S. V. Solomonov, and A. S. Khaikin

A Cryogenic System Containing Liquid Helium for On-Board Radiation
  Receivers.   91
  A. B. Fradkov and V. F. Troitskii

Band-Pass Filters for the Submillimeter Range.   101
  S. V. Solomonov, O. M. Stroganova, and A. S. Khaikin

Properties of the Construction of an On-Board Submillimeter Telescope   111
  V. N. Bakun, P. D. Kalachev, A. E. Salomonovich, and A. S. Khaikin

An On-Board Submillimeter Spectroradiometer.   117
  A. A. Kobzev, V. I. Lapshin, V. F. Troitskii, and A. S. Khaikin

Polarizing Devices for the Submillimeter Range   125
  V. I. Lapshin

Adjusted Deformations of Mirror Systems of Fully Steerable Radio Telescopes. .   137
  P. D. Kalachev, A. N. Kozlov, V. B. Tarasov, and V. N. Titov

Limiting Dimensions of a Fully Steerable Parabolic Mirror for a Radio
  Telescope.   147
  P. D. Kalachev

Experimental Study of Structural Systems of Aerodynamic Compensators in
  Application to Parabolic Antennas.   157
  V. E. D'yachkov, S. L. Myslivets, and V. P. Nazarov

A Parabolic Radio Telescope Antenna with a Radially Balanced Main Mirror . . .   167
    P. D. Kalachev, V. P. Nazarov, I. A. Emel'yanov, V. L. Shubeko,
    and V. B. Khavaev

Study of Elastic Properties of a Fully Steerable Parabolic Antenna for a Radio
    Telescope.   173
    P. D. Kalachev and V. E. D'yachkov

An Automatic Data Processing System for Radio Astronomical Observations . . .   189
    M. V. Konyukov and V. Yu. Bunakov

The Synchronous-Tracking Drive System for the RTI-7.5/250 Radio Telescope
    of the Moscow Technical College.   199
    A. A. Parshchikov and I. A. Emel'yanov

The RTI-7.5/250 Reflecting Radio Telescope with a Fully Steerable Parabolic
    Antenna   205
    P. D. Kalachev, V. P. Nazarov, A. A. Parshchikov, and B. A. Rozanov

A REFLECTING X-RAY TELESCOPE FOR AN ORBITAL ASTROPHYSICAL STATION

I. L. Beigman, L. A. Vainshtein, Yu. P. Voinov, D. A. Goganov,
N. I. Komyak, S. L. Mandel'shtam, I. P. Tindo,
N. A. Shatskii, and A. I. Shurygin

INTRODUCTION

The classical instrument for the study of cosmic sources of x radiation is the photon counter equipped with a mechanical collimating system for the isolation of radiation from individual sections of the celestial sphere. Such "collimator telescopes" have been used since the very inception of x-ray astronomy in 1962 and, in particular, in the especially fruitful experiment on the Uhuru satellite [1]. For a number of years, however, considerable attention has been paid to reflecting x-ray telescopes of "glancing incidence." Such telescopes are especially promising for studies in the soft x-ray region of the spectrum of $\lambda > 10$ Å ($E < 1$ keV). At shorter wavelengths the glancing angle of the ray must be very small ($< 1°$) and mirrors with a focus of several meters are required to obtain a significant effective area. Studies of sources of soft x radiation have already been conducted using reflecting telescopes mounted on rockets [2-5].

The principal advantage of a reflecting telescope is the possibility of concentrating the x radiation from a large area on a detector of small dimensions. This considerably reduces the contribution from the cosmic ray background. Moreover, the detector is shielded from the soft electron component. Finally, an important advantage of a reflecting telescope is the fact that owing to the small area of the entrance window of the counter one can use thinner films, which is especially important in the soft x-ray region where films of even a few microns markedly absorb the radiation. It also becomes possible to use other types of detectors of small area, such as channel multipliers and semiconductor detectors.

Of course, the creation of a reflecting telescope is connected with additional technical difficulties. These difficulties increase considerably if one takes the next step, proceeding to the construction of an x-ray image on a coordinate-sensitive detector. However, in the present case we did not set this task. The smallness of the field of view of a telescope with a paraboloidal mirror also makes additional demands on the accuracy in aiming the telescope at the object.

The RT-4 reflecting x-ray telescope, intended for the study of the soft x radiation of discrete sources using orbital astrophysical stations, is described in the present report. Besides the telescope proper, the apparatus has autonomous systems of control and celestial orientation which are not considered here. A general description of the instrument is given in Section 1 and a detailed description of its separate units in Section 2.

1

Fig. 1. General view of detector assembly of telescope.

1. GENERAL SCHEME OF TELESCOPE

The reflecting x-ray telescope is intended for observations in the soft x-ray region of the spectrum of 40-60 Å. The telescope was developed for installation on orbital stations which provide an observation time of 10 min or more for one source.

The general appearance and a diagram of the telescope are shown in Figs. 1 and 2. The x radiation is focused by the parabolic mirror onto the entrance window of a proportional photon counter. The diaphragm at the mirror entrance prevents the arrival at the counter of direct rays which have not undergone reflection and attenuates the flux of the soft electron component.

The spectral region of sensitivity of the telescope is determined by the transparency of the film of the counter window (2 μ polypropylene).* In addition, the mirror efficiently reflects only radiation with $\lambda > 15$ Å.

---

* The authors are grateful to V. A. Nazarov for active participation in the development of the technology for obtaining thin polypropylene films and to O. D. Lesnykh (Plastpolimer Non-governmental Organization, Leningrad) for the sample of the original unoriented film of high uniformity.

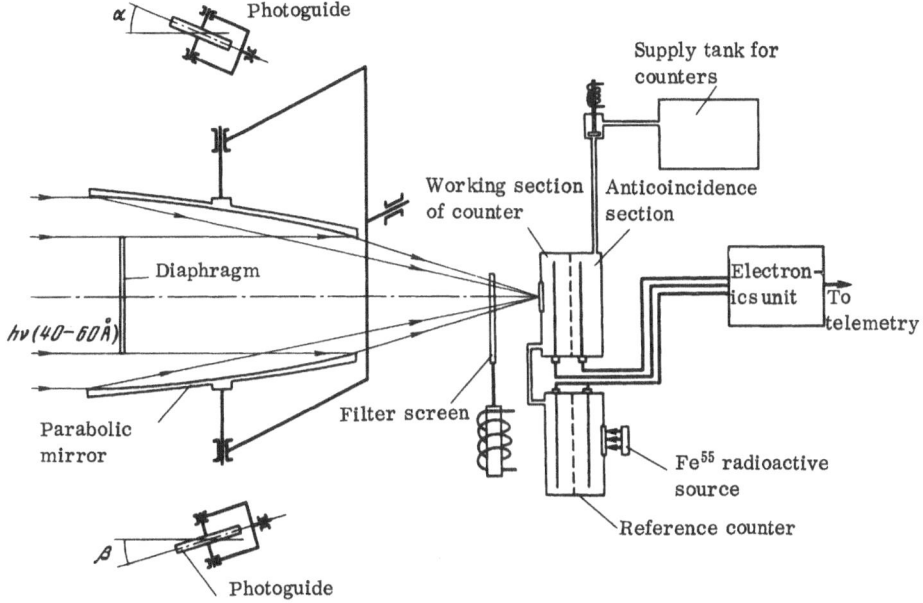

Fig. 2. Diagram of x-ray telescope.

A rotating filter which periodically interrupts the beam of x-ray radiation serves to determine counting level from the penetrating cosmic ray background and the ultraviolet radiation. The filter is made of quartz which is transparent to ultraviolet radiation in the region of $\lambda > 1700$ Å.

The thin plastic film of the counter window does not provide complete air tightness and therefore during an observing session the counters were connected by an electric pressure valve to a tank with a capacity of 3.6 liters containing the gas mixture. Between sessions the counters are disconnected from the tank and a considerable fraction of the gas escapes from the volume of the counters (~ 100 cm³). A system for regulating the high-voltage supply serves to assure the constancy of the coefficient of gas multiplication during the gradual decrease in the gas pressure in the tank.

A description of the separate elements of the telescope is presented in the following section. Here we will dwell only on the data characterizing the sensitivity of the telescope (see Section 2).

The effective area of the mirror for $\lambda = 44$ Å with allowance for the coefficient of reflection from the coating of Ni is $S_x = 100$ cm². The counter efficiency at this wavelength is $\eta_0 = 0.3$. Thus, the telescope efficiency is $\eta_0 S_x = 30$ cm²/photon in the interval $\Delta E = 0.07$ keV.

The minimum flux detected depends on the counting level of the cosmic ray background C_N and on the x-ray background C_x and on the observing time. The use of a mirror permits a decrease in the detector size and consequently in C_N. In our first experiment, however, this fact is not fully utilized: Because of possible errors in aiming the telescope axis at the object studied the size of the detector must be rather great. The diameter of the entrance window of

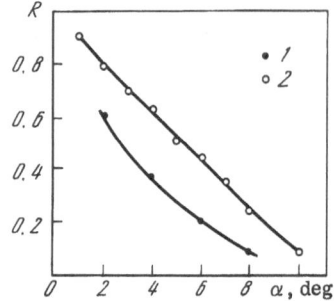

Fig. 3. Reflection coefficient at λ = 44.4 Å. 1) Replica of nickel; 2) optically polished specimen [6].

the counter was chosen as 3 cm (field of view ±1.8°). At present there are insufficient data on the intensity and energy distribution of the cosmic ray background in the region of 0.1-0.6 keV of interest to us. Evidently, for a counter with a 3-cm window one can hope to obtain a value C_N < 1 imp/sec (see Section 2, Part b). Taking an observing time of 500 sec we obtain a value (at the 3σ level) of F_{min}(0.28-0.20 keV) = 6 · 10^{-2} photon/cm^2 · sec, which corresponds to a flux density F_{min} (0.28 keV) = 0.1 photon/cm^2 · sec · keV. The expected counting rate from the diffuse x-ray background in the plane of the Galaxy is C_d ~ 1 imp/sec.

2. PRINCIPAL ELEMENTS OF TELESCOPE

a) Mirror

The telescope mirror consists of a truncated paraboloid. A paraboloid does not satisfy the Abbe sine condition, i.e., it gives a considerable coma for off-axis rays and does not allow one to obtain an image of the object. As is known, the coma can be eliminated by using twofold reflection, as in a combination of paraboloid and hyperboloid. In our case, however, with the relatively large size of the counter window (3 cm) there was no need for this. The use of two surfaces also considerably decreases the effective area.

The mirror has an entrance diameter of 19.7 cm and a distance of 62.4 cm from the focal point to the entrance opening. The shape of the reflecting surface is described by the equation

$$1.52z = x^2 + y^2, \quad z = 18.6 - 62.8$$

(dimensions in centimeters). The glancing angle for axial rays varies from 8.2 to 4°.5.

The experimental data [6, 7] on the reflection coefficients for different materials show that nickel is the optimum coating for the spectral region of λ > 44 Å.

Computer calculations of the effective area S_x for different diameters with a fixed focal distance show that at λ = 44 Å the chosen diameter gives S_x = 144 cm^2 [8]. This is only 20% less than the area for the optimum diameter. At λ = 67 Å S_x = 181 cm^2. For comparison we indicate that the total collecting area (for a reflection coefficient equal to unity) is S_0 = 211 cm^2.

The reflecting surface of the mirror was made by taking a replica* from a matrix of the given paraboloidal shape polished by the standard method. The replica was obtained by electrolytic deposit of a Ni layer ~0.1 mm thick on which copper was then built up to the desired thickness (~1.5 mm). Such mirrors are considerably lighter than glass mirrors. It is easy to prepare several copies from one matrix.

_____

* The technology of obtaining replicas was developed by É. V. Tver'yanovich (VNIIT, Moscow).

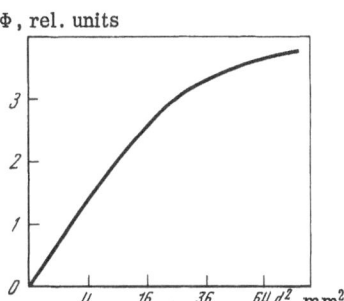

Fig. 4. Dependence of light flux on the square of the diameter of the detector entrance window.

The coefficient of reflection from a surface element of the replica was measured on an x-ray monochromator. The dependence of the reflection coefficient on the glancing angle obtained at $\lambda = 44.4$ Å is presented in Fig. 3 together with the reflection coefficient for optically polished specimens [6]. The somewhat lower reflection coefficient in our case is apparently explained by the insufficient cleanness of the matrix polishing.

Since the diameter of the entrance window of the counter is relatively large, rigid requirements are not placed on the quality of the shape of the generating surface of the mirror. The intensity distribution in the focal plane of the paraboloid for a parallel axial beam, measured in the visible region of the spectrum, is presented in Fig. 4. The diameter of the dispersion spot is 3 mm. Of course, in the x-ray region the dispersion spot may be larger because of the increase in the demands on the quality of polishing of the surface.

The field of view of the telescope was also studied in the visible region. The dependence of the flux on the angle between the principal ray and the optical axis of the mirror obtained is shown in Fig. 5. The angle between the parallel beam and the optical axis of the telescope is laid out along the abscissa. As seen from the figure, the field of view is equal to $\pm 1°.8$. It should be noted that the dispersion spot from a point object is asymmetrical with the maximum in illuminance being displaced toward the axis. This leads to a marked increase in the field of view compared with the field of view determined by the geometrical center of the spot. The latter determination is used in the calculations of [8], which give a field of $\pm 1°.3$ for our case. A calculation of the aberrations made on a computer [8] confirmed the experimental result presented above.

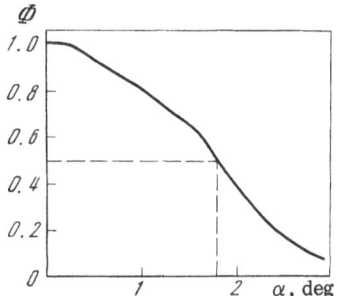

Fig. 5. Dependence of light flux on angle between ray and optical axis of mirror.

b)   Photon Counter

A proportional counter of the SRPP-36 type is used in the telescope. The counter has the form of a short cylinder divided into two parallel sections and the anode filaments are located along the diameter of each section (Fig. 6).

The sections of the counter are divided by beryllium foil 0.1 mm thick. The upper section is used to record the soft x radiation. The lower section is connected to the instrument

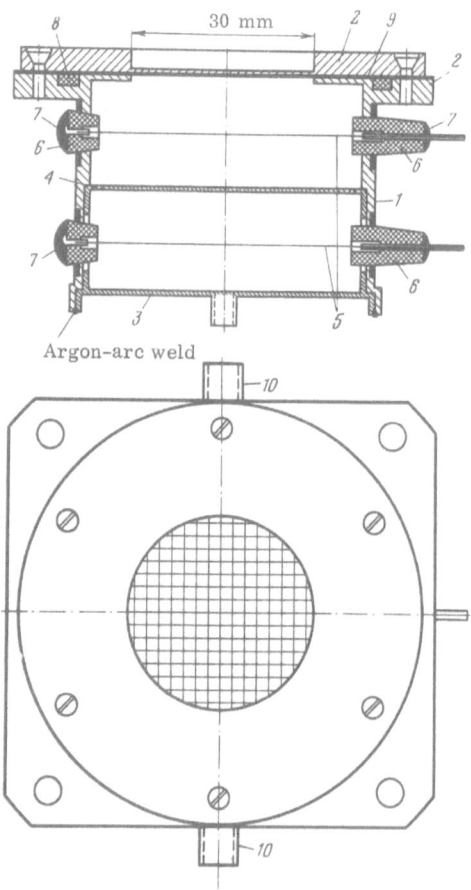

Fig. 6. SRPP-36 proportional x-ray counter. 1) Body (AMts alloys); 2) window grid (U8 steel); 3) bottom (AMts alloys); 4) diaphragm (0.1 mm beryllium);   5) anode "filaments"; 6) ceramic insulators; 7) adhesive joint; 8) rubber gasket; 9) window (2-μ polypropylene); 10) inlet and outlet fittings.

through an anticoincidence circuit with the upper section and serves to decrease the counting rate from the cosmic ray background.

The upper section has an entrance window covered by a 2-μ polypropylene film supported from the outside by a steel grid. The diameter of the window is 3 cm. The counter is filled with a gas mixture of 90% Ar + 10% CH_4 to a pressure of 1-2 atm.

The escape of the gas through microleaks occurs with the use of a thin plastic film. A mode of operation in individual sessions with intervals of a day or more between them is planned for this telescope. During the time between sessions the gas escapes almost completely from the volume of the counter. Therefore between sessions the counter is disconnected from the tank containing the mas mixture and is joined with it only during a session. The construction of the system of gas supply is presented in Fig. 7. Through the careful selection of the film specimens it is possible to assure a flow of no more than 10^{-2} cm^3/min at a pressure drop of 1 atm. In this case the loss during a session is only a small fraction of the counter volume.

The film of the entrance window selects a spectral interval of 0.28-0.20 keV (44-60 Å). To reduce the sensitivity to ultraviolet radiation the inner surface of the counter is covered with a layer of graphite and a layer of carbon of 40 $\mu g/cm^2$ is deposited on the outer surface of the entrance window film. A thin layer of aluminum is deposited on the inner surface to provide conductivity. The transparency of such a film to radiation in the 1800 Å region does not exceed 10^{-3}.

The coefficient of gas multiplication is about 10^4. The counter operates in the proportional mode with the recording of pulses in the energy range of 0.6-0.1 keV.

The amplitude distribution during monochromatic irradiation with an Fe[55] source (2.09 Å) is shown in Fig. 8. The halfwidth of the distribution is 25%.

The proportionality of the average pulse amplitude to the photon energy was tested in the range of 0.36-6 keV. The presence of proportionality makes it possible to use for adjustment and testing of the instrument the radiation of an Fe[55] source, work with which is possible at normal atmospheric pressure in contrast to the working spectral range around 44 Å.

The calculated efficiency η_0 of the counter is determined by the transparency τ_0 of the window before the K-edge of absorption (E_0 = 0.28 keV) and by the effective energy range

$$\Delta E = E_0 [1 - (1 + 1/\tau_0)^{-1/3}],$$

where τ_0 is the optical thickness of the film. It is assumed that $\tau(E) = \tau_0 (E_0/E)^3$. The value $\eta_0 = Ke^{-\tau_0}$ where K = 0.5-0.7 is the transparency of the grid. For a film of 2-μ polypropylene + 40 $\mu g/cm^2$ C the calculated values are τ_0 = 0.7, η_0 = 0.30, and ΔE = 0.07 keV.

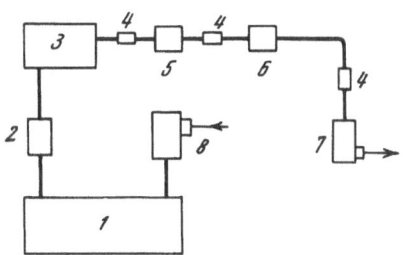

Fig. 7. Gas supply system. 1) Tank containing quenching mixture of 90% Ar + 10% CH_4; 2) filter; 3) impulse electric valve which opens automatically when instrument power is turned on; 4) rubber connecting joints; 5) working counter; 6) reference counter; 7) drain valve; 8) charging valve.

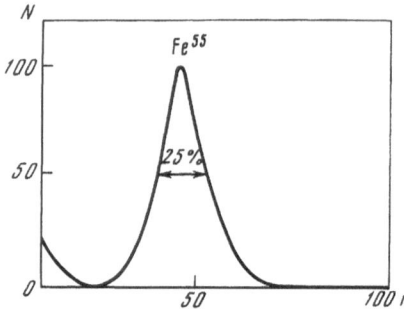

Fig. 8. Amplitude distribution of pulses during irradiation with Fe^{55} source. Along abscissa: channel numbers of pulse analyzer (proportional to amplitude of pulses).

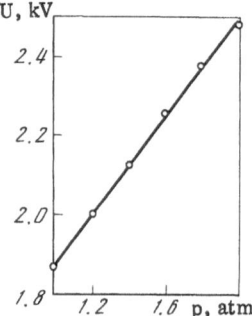

Fig. 9. Dependence of working voltage of telescope counter on pressure of gas mixture.

The measurement of η_0 was conducted with the isolation of a narrow interval around 0.28 keV from the spectrum of an x-ray tube using a filter of thick Dacron film. A Lukirskii counter with a known absolute sensitivity [10] served as the standard. The value of η_0 obtained is close to the calculated value. Thus, the total efficiency of the telescope is $S_x \eta_0 \Delta E \geq 2 \text{ cm}^2 \cdot \text{keV}$.

The dependence of the working voltage of the telescope counter on the gas pressure is shown in Fig. 9. The voltage is regulated automatically using a "reference" counter of similar construction which is constantly irradiated by an Fe^{55} source (see Part c).

c)   Electrical Diagram of X-Ray Telescope

The principal elements of the telescope diagram are:

1. a circuit for measuring the pulse counting rate;

2. a device for regulating the voltage of the power supply to the proportional counter;

3. a circuit for controlling the filter actuator and blocking the inputs of the working and control channels;

4. a stabilized power supply unit;

5. a circuit for controlling the pulsed electric pressure valve;

6. a device for indicating the position of the filter and the electric pressure valve, a pressure pickup for the gas in the tank, and a temperature pickup.

A structural diagram of the telescope is presented in Fig. 10.

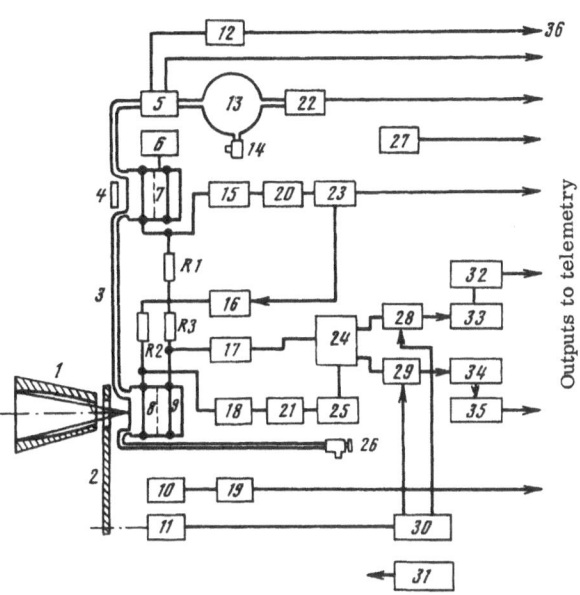

Fig. 10. Structural diagram of telescope. 1) Telescope
mirror; 2) rotating filter; 3) gas line; 4) radioactive source;
5) electric pressure valve; 6) bias voltage source; 7) reference counter; 8) x-ray section of working counter; 9) control section of working counter; 10) photodetector of filter
position; 11) filter actuator; 12) control circuit for electric
pressure valve; 13) tank containing gas mixture; 14) charging
valve; 15) input stage of amplifier for reference counter; 16)
regulating high-voltage transformer; 17) input stage of amplifier for control section; 18) input stage of amplifier for
x-ray section; 19) amplifier of photodetector of filter position; 20) amplifier stage for reference counter; 21) amplifier stage for x-ray section; 22) pressure pickup; 23) high-
voltage regulating unit; 24) anticoincidence and discrimina-
tion circuit; 25) pulse former; 26) drain valve; 27) temperature pickup; 28, 29) blocking switches; 30) unit for control
of filter actuator and blocking of inputs; 31) transformer for
stabilized power supply; 32) summation device; 33, 34) counting circuit; 35) summation device; 36) valve control commands.

The measuring circuit contains pulse amplifiers for the working 18, 25 and control 17 sections of the counter, a pulse selector unit 24, counting circuits 33, 34, and "summation devices" 32, 35 which transform the state of the counting elements into analog form.

A current pulse amplifier with a gain $K = 5 \cdot 10^4$ is used to amplify the pulses of the proportional counter. The change in the amplifier gain with a change from -60 to $+60°C$ in the temperature of the surrounding medium does not exceed $\pm 5\%$. The internal noise of the amplifier is $3 \cdot 10^{-15}$ C.

The pulse selection unit consists of an amplitude selector and an anticoincidence circuit. The selection unit assures that the coounting circuit of the working channel records pulses from the working section of the counter which correspond to photons in the spectral interval of 0.1-0.6 keV and do not coincide in time with pulses of the control section of the counter.

The counting circuit of the control channel records pulses of the control section of the counter corresponding to an energy release of more than 0.3 keV, pulses of the working section with an amplitude of more than 0.6 keV, and pulses of the working section exceeding the 0.1 keV level and coinciding in time with pulses of the control section.

The amplitude selection of pulses is accomplished with a circuit based on tunnel diodes. The anticoincidence circuit operates on the basis of delayed blocking generators.

The amplifier circuit of the working section of the counter includes a sensitivity switch 21 permitting a 50-fold decrease in the gain. This switch is used in laboratory adjustment of the recording circuit. In the mode of low sensitivity the pulses from the Fe^{55} source ($E = 5.94$ keV) are at the lower boundary of the recorded amplitudes of the working channel.

The counting circuits of the working and control channels each contain four double triggers and operate in a continuous counting mode. The conversion of the state of the counting elements into analog form is performed by summing the currents of each trigger with allowance for its weighting coefficient in general information. A voltage proportional to the total current of the triggers is fed to the input of the telemetry system. This voltage changes stepwise following the arrival of the next pulse at the input of the counting circuit. With a high-interrogation telemetry system this type of conversion makes it possible to record the arrival of each pulse from the counter.

Device for Regulating Voltage of Power Supply to Proportional Counter. During the operation of the instrument the gas mixture in the tank is gradually exhausted and the pressure decreases, which requires a corresponding change in the voltage of the counter's power supply. The regulating device used for this purpose includes reference source 4 which is isotope Fe^{55}, the proportional counter 7, the pulse amplifier 15, the high-voltage regulating unit 23, the high-voltage transformer 16, and the bias voltage transformer 6. The voltage regulating device operates on the principle of a slave system, maintaining a constant amplitude in the pulses of the counter irradiated by the reference source.

The principal and reference counters have a common gas supply from the tank, i.e., they operate at the same pressure. The reference counter is constantly irradiated by the radioactive source. The amplified pulses of the reference counter (counting rate $\sim 10^3$ imp/sec) are supplied to the input of the high-voltage regulating unit where their amplitude is compared with a reference voltage. If the amplitude of the pulses exceeds the reference voltage then the voltage at the output of the high-voltage transformer of the counter's power supply is reduced. An RC filter with a time constant of ~ 30 sec is included in the power supply circuit of the counters.

Under laboratory conditions the voltage regulating device permits the amplitude of the counter pulses to be kopt within the limits of $\pm 10\%$ during a change in the pressure of the gas mixture from 2.6 to 0.9 atm and during variation in the temperature of mass of the detector assembly within limits of ±60°C. Interference from the recording of the primary cosmic ray background by the reference channel reduces the upper limit of allowable pressure to 1.6 atm.

The circuit 30 for controlling the filter actuator and blocking the inputs of the working and control channels produces the voltage needed to supply the solenoid with which the filter is moved and puts out signals to block the inputs of the counting circuits (using the diode switches 28 and 29) while the filter is changed from one position to the other. Blocking the inputs prevents disturbances in the counting circuits during changing of the filter.

The circuit consists of a generator of one-second pulses, counting circuits which determine the time of the open and closed positions of the filter, and an amplifier stage. This provides alternate exposure without the filter for ~40 sec and with the quartz filter in front of the entrance window of the main counter for ~10 sec.

The stabilized power supply unit 31 contains an input voltage stabilizer, a two-cycle transformer, and an output voltage stabilizer. The power supply unit is designed to operate from an input voltage of 27^{+7}_{-4} V and provides the following output voltages for the supply of the telescope circuits: 6.3 V ± 5% and 1.2 V ± 5% to supply the counting circuits; 12 V ± 1% to supply the amplifier stages; 13.5 V ± 5% to supply the high-voltage transformer. In addition, the power supply unit provides the galvanic uncoupling of the primary power supply from the electrical recording circuit.

The circuit 12 for controlling the pulsed electric pressure valve is constructed on the basis of a univibrator which operates when the power is turned on. It supplies a pulsed voltage of 27^{+7}_{-4} V amplitude for 0.1 sec to the winding of the electric valve (resistance 30 Ω).

The device for indicating the position of the filter consists of two miniature incandescent lamps as light sources, silicon photodiodes, and amplifier stages. The output signals of the device provide information on the position of the filter.

The free contacts of the electric pressure valve are used to indicate the position of the valve.

In conclusion, the authors consider it their pleasant duty to thank V. I. Shurygin, V. A. Slemzin, V. I. Svirin, V. A. Drozdovskii, A. I. Parshin, and V. A. Nazarov for help in fabricating the instrument.

LITERATURE CITED

1. R. Giacconi, E. Kellogg, P. Gorenstein, H. Gursky, and H. Tananbaum, Astrophys. Lett., 165, 27 (1971).
2. P. Gorenstein, B. Harris, H. Gursky, R. Giacconi, R. Novick, and P. Vanden Bout, Science, 172, 369 (1971).
3. D. J. Yentis, R. Novick, and P. Vanden Bout, Astrophys. J., 177, Part 1, 365, 375 (1972).
4. D. J. Yentis, J. R. P. Angel, D. Mitchell, R. Novick, and P. Vanden Bout, New Techniques in Space Astronomy (IAU Symposium 41) (1971), p. 145.
5. P. Gorenstein, A. De Caprio, R. Chase, and B. Harris, Rev. Sci. Instr., 44, 539 (1973).
6. A. P. Lukirskii, E. P. Savinov, O. A. Ershov, and Yu. F. Shepelev, Opt. i Spektr., 16, 310 (1964).
7. O. A. Ershov, I. A. Brytov, and A. P. Lukirskii, Opt. i Spektr., 22, 127 (1967).
8. I. L. Beigman, L. A. Vainshtein, Yu. P. Voinov, and V. P. Shevel'ko, Trudy Fiz. Inst. Akad. Nauk SSSR, 77, 14 (1974).

9.   R. Giacconi, W. P. Reidy, G. S. Vaiana, L. P. Vanspeybroeck, and T. F. Zehnpfennig, Space Sci. Rev., 9, 3 (1969).

10. A. P. Lukirskii, I. A. Brytov, and O. A. Ershov, Izv. Akad. Nauk SSSR, Ser. Fiz., 27, 446 (1963).

MIRROR SYSTEMS FOR X-RAY TELESCOPES

I. L. Beigman, L. A. Vainshtein, Yu. P. Voinov, and V. P. Shevel'ko

INTRODUCTION

As is known, studies of the soft x radiation of cosmic objects using proportional flow counters of large area are associated with considerable technical difficulties. At the same time, the use of reflecting x-ray optics is possible in this region of the spectrum (softer than 1 keV). It is advisable to examine the use of reflecting x-ray telescopes in two aspects. On the one hand, the reflecting optics makes it possible to have a comparatively large collecting area with small detector dimensions, which decreases the technical difficulties to a considerable extent and increases the signal-to-noise ratio; on the other hand, reflecting objectives open up the possibility in principle of obtaining an image analogous to that provided by telescopes in the visible region of the spectrum.

The first fabrication and use of an instrument with reflecting collecting optics (a parabolic mirror) for studies in the soft x-ray region of the spectrum are discussed in [1, 2]. An instrument of similar type designed for mounting on orbital stations is described in an article [3] of the present collection.

A large number of works have been devoted to x-ray mirror systems providing an image in the focal plane. Wolter [4, 5] showed that an even number of reflections is a necessary condition for obtaining an image. A detailed analysis of systems of a confocal paraboloid and an extensive bibliography are contained in the review [6] and in [7]. An x-ray photograph of a solar flare with a resolution of ~2" was obtained [8] using a system of this type. A system with crossed parabolic mirrors is described and studied in [9].

The purpose of the present work is the study of collecting systems (cone, paraboloid) and objectives (the combination of a paraboloid with a hyperboloid) from the point of view of obtaining the greatest effective area for the given overall size of the system. Since an x-ray mirror system in itself represents a tube for which the diameter is considerably less than the length, the main overall dimension is considered to be the length, i.e., the focus of the system. The combination of a paraboloid with a hyperboloid also produces a distribution of intensity in the focal plane as a function of the angle between the parallel bundle of rays and the axis of the system. Such a distribution makes it possible to judge the aberrations and resolving power of the objectives.

The small size of the entrance window of the detector used in work with reflecting systems opens up possibilities for the use of semiconductor detectors and channelotrons in addition to gas counters. The prospects for the use of these detectors are examined in Section 6.

13

1. REFLECTION COEFFICIENT

Modern reflecting x-ray optics is based on the phenomenon of total "external" reflection. X-ray beams can be efficiently reflected from surfaces because the index of refraction of the material is less than unity for x-rays. The index of refraction can be represented in the form

$$n = 1 - \delta - i\beta. \tag{1}$$

For the x-ray region of the spectrum in the transition from air into the material the values δ and β are positive and small in absolute value. Since the real part of the index of refraction is less than unity there exists an angle of total "external" reflection determined by the condition

$$\cos \alpha_0 = 1 - \delta, \tag{2}$$

where α_0 is the angle between the ray the reflecting surface.

For small δ the condition (2) can be rewritten in the form

$$\alpha_0 = \sqrt{2\delta}. \tag{2a}$$

If the imaginary part β of the index of refraction, which is responsible for absorption, were equal to zero then for angles less than α_0 the reflection coefficient K would be equal to unity. The presence of absorption leads to a marked decrease in reflection at angles less than α_0. The reflection coefficient with allowance for absorption is given by the Fresnel equations. At small angles they have the form

$$K(\alpha) = \frac{(\alpha - \theta_1)^2 + \theta_2^2}{(\alpha + \theta_1)^2 + \theta_2^2},$$

$$\theta_1 = \frac{1}{\sqrt{2}} \left[\sqrt{(\alpha^2 - 2\delta)^2 + 4\beta^2} + (\alpha^2 - 2\delta) \right]^{1/2}, \tag{3}$$

$$\theta_2 = \frac{1}{\sqrt{2}} \left[\sqrt{(\alpha^2 - 2\delta)^2 + 4\beta^2} - (\alpha^2 - 2\delta) \right]^{1/2}.$$

Durable coatings for which the reflection coefficient has a considerable value at comparatively large glancing angles are of particular interest for x-ray mirrors.

Systematic measurements of the coefficient of reflection from different coatings as a function of the angle of incidence and the wavelength of the incident x radiation have been made by Ershov, Brytov, and Lukirskii [10] and in [11]. It follows from the data presented in [10, 11] that nickel and gold are the most promising materials for the coatings of x-ray mirrors. The reflection coefficients of these elements for different glancing angles are presented in Figs. 1 and 2. The parameters δ and β, taken from [10], are given in Table 1.

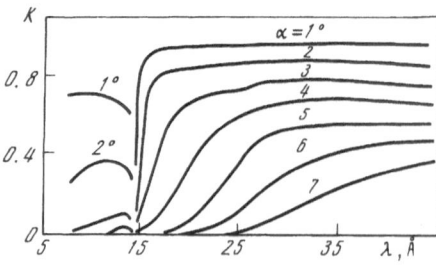

Fig. 1. Dependence of reflection coefficient of nickel on wavelength.

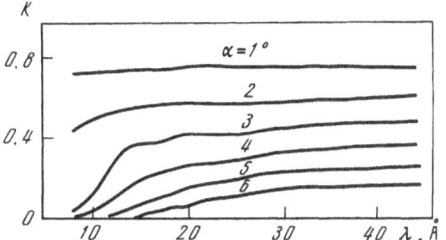

Fig. 2. Dependence of reflection coefficient of
gold on wavelength.

TABLE 1

λ, Å	β/δ	δ·10⁴	δ/λ²	β·10⁴
		Nickel		
8.34	0.32±0.03	5.91±0.17	8.48	1.9±0.2
9.89	0.40±0.04	8.00±0.20	8.17	3.2±0.3
10.44	0.50±0.05	8.70±0.20	8.15	4.4±0.4
12.25	0.60±0.06	8.85±0.21	5.90	5.3±0.5
13.34	0.70±0.07	10.20±0.22	5.72	7.2±0.7
14.56	7.00±0.70	1.37±0.08	0.65	9.6±1.0
15.97	0.17±0.02	12.30±0.20	4.83	2.1±0.2
17.59	0.16±0.02	17.50±0.30	5.65	2.8±0.3
19.4	0.17±0.02	21.70±0.30	5.74	3.7±0.4
21.64	0.20±0.02	30.50±0.40	6.50	6.1±0.6
23.62	0.22±0.02	33.30±0.40	5.98	7.3±0.7
24.78	0.23±0.02	39.40±0.44	6.43	9.1±0.9
27.42	0.21±0.02	47.30±0.48	6.30	10.0±1.0
31.36	0.25±0.03	60.00±0.50	6.12	15.0±1.5
		Gold		
8.34	0.34±0.03	9.15±0.21	13.10	3.1±0.3
9.89	0.38±0.04	11.50±0.24	11.75	4.4±0.4
12.254	0.52±0.05	18.00±0.30	12.00	9.4±0.9
13.34	0.58±0.06	21.20±0.30	11.90	12.3±1.2
14.56	0.65±0.07	24.10±0.30	11.40	15.7±1.6
15.97	0.70±0.07	26.60±0.40	10.45	18.6±1.9
17.59	0.80±0.08	29.20±0.40	9.45	23.4±2.3
19.45	0.80±0.08	33.50±0.40	8.85	26.8±2.7
21.64	0.82±0.08	34.80±0.40	7.43	28.5±2.9
23.62	1.00±0.10	37.80±0.40	6.75	37.6±3.8
24.78	1.40±0.10	40.00±0.40	6.50	56.0±5.6
27.42	1.50±0.20	40.80±0.50	5.44	61.0±6.0
31.36	1.30±0.10	47.50±0.50	4.85	61.5±6.2

The values of δ and β for nickel at the wavelength 44.4 A, which are used in the subsequent calculations, were determined by trial and error on a computer from the known reflection coefficient (Fig. 1) for three glancing angles: 4, 7, and 8°. The values $\delta = 1.16 \cdot 10^{-2}$ and $\beta = 3.23 \cdot 10^{-3}$ were obtained.

It should be noted that the reflection coefficient in the x-ray region depends strongly on the quality and technology of the preparation of the surface and special control is needed.

2. COLLECTING MIRROR

a) Cone

The cone serves as the simplest collecting mirror for x-ray astronomy (Fig. 3). It makes it possible to increase the flux density of the radiation parallel to the main optical axis

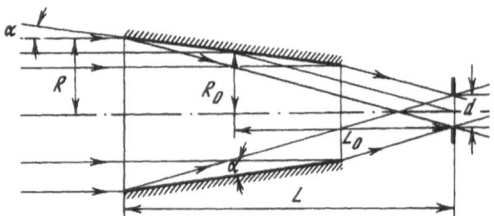

Fig. 3. Diagram of a reflecting cone.

by

$$\frac{8R_0 \cos 2\alpha}{d} K(\alpha) \text{ times,} \tag{4}$$

where d is the diameter of the counter window, $K(\alpha)$ is the reflection coefficient at the given angle α, and R_0 is the mean radius of the cone.

It is easy to show that in this case the ratio signal/(noise)$^{1/2}$ does not depend on the diameter of the counter window.

In addition to its principal task, to gather the flux, such a cone simultaneously serves as a collimator with an angle of view (halfwidth) of $\Delta\alpha \approx \pm 0.4d/L_0$, where L_0 is the distance from the counter window to the middle of the working section of the cone.

The effective area of an x-ray counter with a collecting cone is

$$S_{ef} = 2\pi d^2 \cos 2\alpha \cdot \cos \alpha \left(2\frac{L}{d}\sin \alpha - \cos \alpha\right) K(\alpha), \tag{5}$$

where L is the distance from the counter window to the entrance pupil.

The minimum angle α_m at which a parallel beam of rays after reflection still completely fills the detector window is determined by the condition

$$\alpha_m = \frac{d}{L}. \tag{6}$$

One usually works at angles $\alpha > \alpha_m$. In this case

$$S_{ef} \approx 4\pi L d [\alpha K(\alpha)]. \tag{7}$$

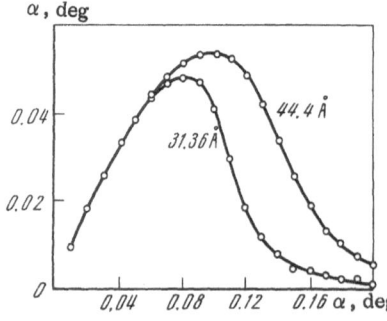

Fig. 4. Dependence on glancing angle α of effective area $S_{ef} \sim \alpha K(\alpha)$ of a cone made of nickel.

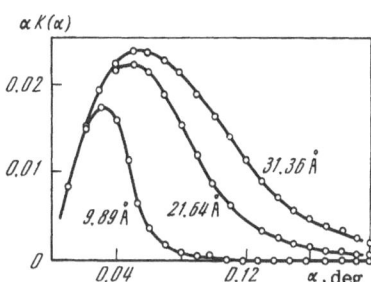

Fig. 5. Dependence on glancing angle α of effective area $S_{eff} \sim \alpha K(\alpha)$ of a cone made of gold.

The optimum dimensions R of the cone at a given length L, counter diameter d, and known dependence $K = K(\alpha)$ of the reflection coefficient on the glancing angle [see Eq. (3)] for a given wavelength are determined from the condition of maximizing the effective area. The function $\alpha K(\alpha)$ for Ni (λ = 44.4 and 31.36 Å) and Au (λ = 9.89, 21.64, and 31.36 Å) is presented in Figs. 4 and 5.

b) Paraboloid

A considerably greater gain in signal-to-noise ratio is achieved with the help of paraboloidal mirrors. In this case the magnitude of the useful signal is determined by the effective area of the collecting mirror while the noise is determined by the detector dimensions, which in principle can be made sufficiently small.

A diagram of the mirror-paraboloid and the designations are presented in Fig. 6.

The equation for the paraboloid in these designations has the form

$$x^2 + y^2 = 2p(z_p - z), \tag{8}$$

where the z axis coincides with the axis of symmetry of the system and is the main optical axis. The parameter p of the paraboloid and the distance z_p from the apex to the entrance window of the mirror are

$$p = -L_p + \sqrt{L_p^2 + b^2},$$

$$z_p = \frac{b^2}{2p}, \tag{9}$$

where b is the radius of the entrance pupil and L_p is the "focal" distance.

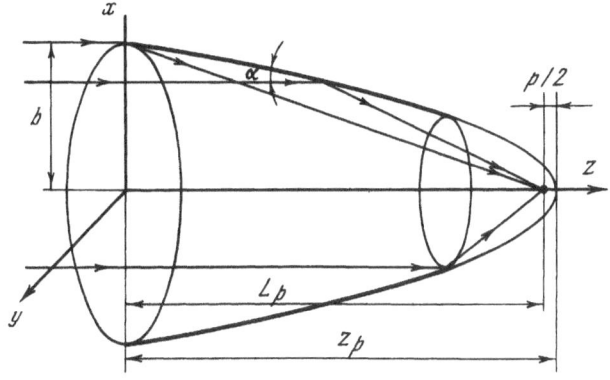

Fig. 6. Diagram of mirror-paraboloid.

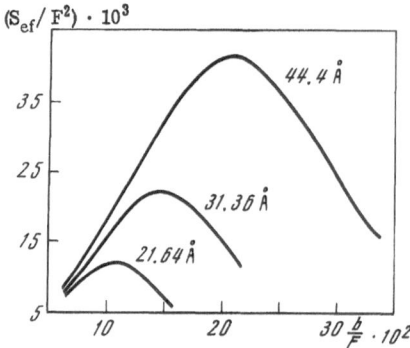

Fig. 7. Dependence of effective area on entrance radius b and focus F of a paraboloid
made of nickel.

Fig. 8. Dependence of effective area on parameters of a paraboloid made of gold.

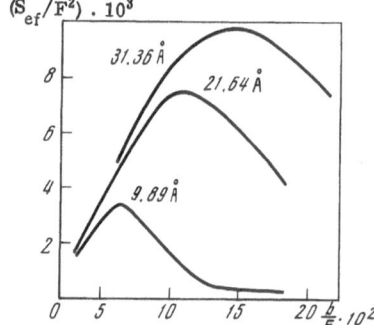

The effective area of a paraboloidal mirror is

$$S_{ef} = 2\pi \int_0^b \rho K(\alpha)\, d\rho, \qquad (10)$$

where $\rho = \sqrt{x^2 + y^2}$, while the reflection coefficient is determined in accordance with Eqs. (3).

The dependence of the effective area on the parameters (b and $L_p = F$) of a paraboloid
with a coating of Ni (for λ = 44.4, 31.36, and 21.64 Å) and gold (for λ = 9.89, 21.64, and 31.36 Å)
is presented in Figs. 7 and 8. It is seen that it is advisable to use a paraboloid with a nickel
coating for work in the range of wavelengths greater than 20 Å. For the range of 10–20 Å
a greater effective area can be obtained by a gold coating. Calculations conducted for a coating
of aluminum show that it gives a considerably lower effective area in the indicated ranges.
For the region $\lambda < 10$ Å it is technically impractical to produce reflecting collecting optics
with the materials known at present.

The dependence of the effective area of the paraboloid on the angle between a parallel
beam and the axis is determined by the detector dimensions. As an illustration this dependence,
obtained through calculations, is presented in Fig. 9 for a paraboloid with a nickel coating at
a wavelength of 44 Å. The size of the field of view $\Delta\alpha$ is approximately proportional to the

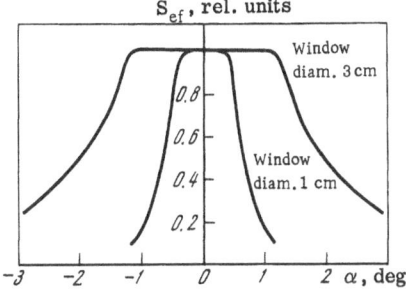

S$_{ef}$, rel. units

Window diam. 3 cm

Window diam. 1 cm

α, deg

Fig. 9. Calculated dependence of S$_{eff}$ (in rel. units) on the angle between a parallel beam (λ = 44 Å) and the axis. Nickel paraboloid with F = 65 cm and b = 12 cm.

detector dimensions: $\Delta\alpha \sim d/F$. Detailed equations for calculations of the ray path are presented in the Appendix.

3. EXPERIMENTAL PARAMETERS OF PARABOLIC MIRROR
OF X-RAY TELESCOPE FOR RANGE OF 44-60 Å

Paraboloidal mirrors of epoxy resin with an aluminum coating and metallic replicas with a nickel coating were fabricated for the soft x-ray region of the spectrum of 44-60 Å.

a) Epoxy Parabolic Mirror

The paraboloidal mirror of epoxy resin was obtained by the method of rotation of a solidifying solution (rotation rate n ≈ 343 rpm) and had the dimensions b = 9.8 cm and F = 63 cm and the parameter p = 0.76 cm.

The quality of the shape of the surface (the correspondence to the equation $x^2 + y^2 = 2pz$) was tested in the visible range using a collimator. The focal spot sizes obtained lie at the limits of divergence of the collimator (~4.5'). Thus, the deviation of the normal to the surface from the assigned value does not exceed ~1'.

Aluminum was chosen as the coating from considerations of simplicity of the technology of vacuum deposition with a comparatively high reflection coefficient in the region of 44 Å.

Measurements of the angular dependence of the reflection coefficient were made on an x-ray crystal monochromator for a specimen cut from a fabricated epoxy mirror. An x-ray tube with an anticathode of carbon was used as the radiation source. The working wavelength — the characteristic line C$_{K\alpha}$ (44.4 Å) with a constant of ~10 — was isolated with a flat crystal of barium stearate (lattice constant d = 50 Å). A proportional flow counter was used as the detector. A film of Zapon served as the window. A comparison between the experimentally

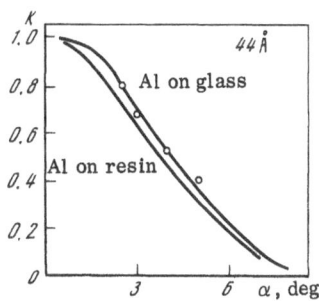

Fig. 10. Dependence of reflection coefficient of aluminum deposited on a paraboloid of resin (lower curve) on the glancing angle. Points are Lukirskii's measurements [11].

K

44 Å

Al on glass

Al on resin

α, deg

measured reflection coefficient and the data of [10], in which aluminum deposited on polished glass was studied, is presented in Fig. 10. The control measurements of the reflection coefficient for aluminum on glass coincide with the results of [10].

b) Metallic Replicas

The metallic paraboloidal mirror of nickel was fabricated by the method of electrolytic deposit on a matrix. A matrix with the optimum dimensions (F = 65 cm) for λ = 44 Å calculated by the method indicated above was turned on a lathe with programmed control and polished by the usual methods. The first (working) layer of the copy deposited is ~0.1 mm of nickel and then comes a layer of copper of the necessary thickness — usually 1.5-2 mm is sufficient.

Mirrors fabricated by this method have a number of advantages over epoxy and glass mirrors. They are considerably lighter and do not require a complicated mounting for reinforcement and heat resistance. Moreover, the method of taking copies makes it possible to obtain a large number of mirrors from one matrix, which is important for work on satellites and rockets when the instruments are not recoverable.

The focusing properties of the surface, which depend on the accuracy in fabricating the paraboloid, were tested in the visible range. For this a parallel beam of rays (divergence ~10^{-3} rad) from a collimator was directed at the mirror along the main optical axis. An iris diaphragm was placed at the focus. The emerging radiation was recorded by a photomultiplier. The resulting distribution of radiation intensity at the focus obtained is represented graphically in Fig. 11. The diameter of the focal spot was 3 mm. We note that the size of the focal spot for an ideal paraboloid would be 0.7 mm for the given divergence.

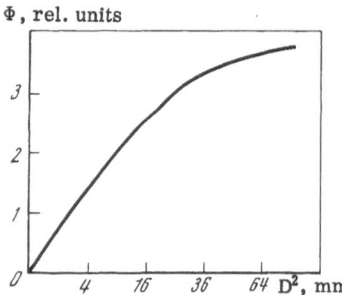

Fig. 11. Experimental distribution of intensity in focal plane of a paraboloid made of nickel.

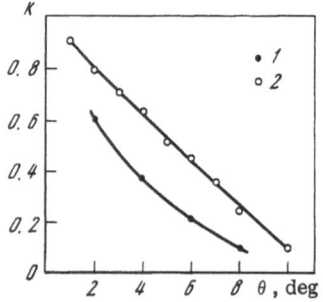

Fig. 12. Dependence of reflection coefficient of glancing angle. 1) For replica made of nickel (λ = 44 Å); 2) optically polished specimen made of nickel [10].

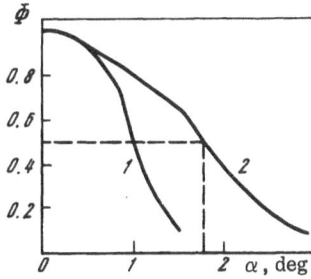

Fig. 13. Dependence of light flux Φ in focal plane on angle between parallel beam and axis. Nickel paraboloid, b = 10 cm, F = 62 cm. 1) Window diameter 10 mm; 2) 30 mm.

To obtain higher accuracies in the fabrication of the mirrors it is necessary to apply special methods of control of the shape of the surface when polishing the matrix. The quality of the polishing also determines the value of the reflection coefficient in the x-ray region of the spectrum. The reflection coefficient was measured by the method described in Section 3, Part a. The results of the experimental measurement together with the data of [10] are presented in Fig. 12.

The field of view of the paraboloidal mirror was also determined in the visible region. The results of the measurements, presented in Fig. 13, for detectors with entrance windows 10-30 mm in diameter are close to the theoretical results (see Fig. 9).

4. OBJECTIVES

A paraboloid has large aberrations of the coma type. An exact calculation shows that for off-axis rays the angular size of the aberrations exceeds the angle between these rays and the axis. This makes it impossible to obtain an image with a paraboloid.

In an optical system which forms an image without a coma the Abbe sine condition must be satisfied, i.e., the ratio $\sin\alpha/\sin\alpha'$ must be constant for all rays, where α and α' are the angles between a given ray and the line joining a point in the object plane and the image plane. As seen from Fig. 14, the Abbe sine condition, grossly violated for a paraboloid, is qualitatively satisfied for a paraboloid with a hyperboloid, since in the latter case α' decreases as α increases.

Wolter showed [4, 5] that the sine condition can be satisfied only with an even number of reflecting surfaces. An objective consisting of a coaxial and confocal paraboloid and hyperboloid is the most promising in this respect. A diagram and the designations for such an objective are presented in Fig. 15.

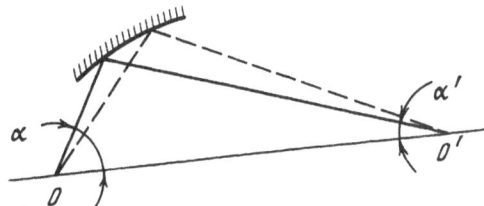

Fig. 14. Illustration of sine condition for a paraboloid.

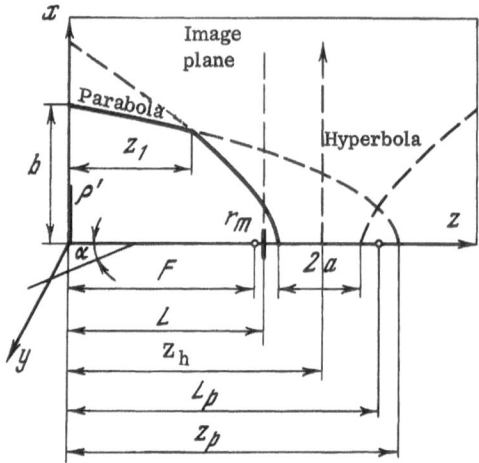

Fig. 15. Diagram of paraboloid + hyperboloid objective.

The equation for the paraboloid in these designations is

$$x^2 + y^2 = 2p(z_p - z),\tag{11}$$

and for the hyperboloid

$$-\frac{x^2 + y^2}{c^2} + \frac{(z - z_h)^2}{a^2} = 1.\tag{12}$$

The z axis coincides with the axis of symmetry of the system.

The condition for the paraboloid and hyperboloid to be confocal is

$$\frac{L_p - F}{2} = \sqrt{a^2 + c^2}, \quad \frac{L_p + F}{2} = z_h,\tag{13}$$

where L_p is the "focal" distance of the paraboloid, F is the focal distance of the system, and z_h is the distance from the entrance pupil to the center of the hyperboloid.

Detailed equations for the calculation of the ray path and the parameters of the paraboloid and hyperboloid are given in the Appendix.

The effective areas calculated with the reflection coefficients from [10, 11] for an objective made of Ni (λ = 44.4 and 31.36 Å) and of Au (λ = 9.89, 21.61, and 31.36 Å) are presented in Table 2.

The dimensions b and L_p, the radius of the entrance pupil and the "focal" distance of the paraboloid, are given in centimeters. The effective area is in square centimeters. The first row of the table (L_p = 65) corresponds to a parabola alone. The underlined numbers correspond to the most effective area for a parabola and for an objective of a paraboloid and a hyperboloid.

Principal attention was paid to the calculation of aberrations. Since the coma for the objective is partly removed its aberration is considerably smaller. The nature and size of the

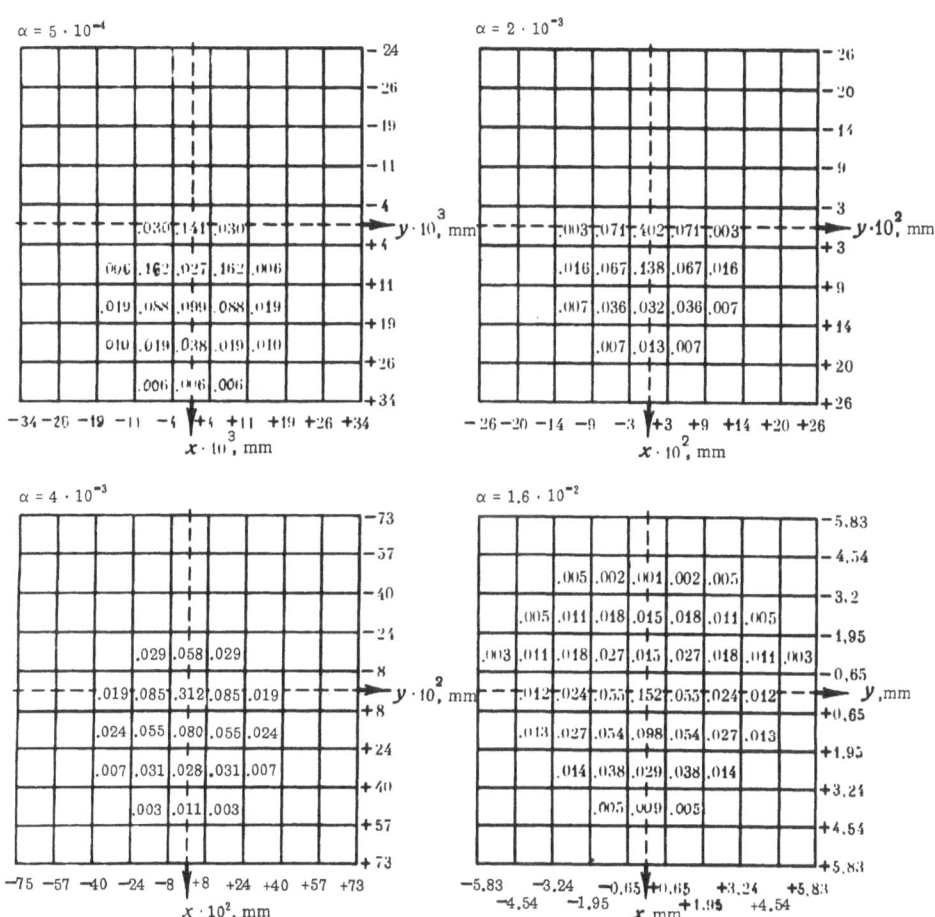

Fig. 16. Distribution of intensity in focal plane for angle α between parallel beam (44.4 Å) and axis of system. Objective (parabola + hyperbola): Ni, b = 16 cm, L_p = 70 cm, F = 65 cm.

TABLE 2

L_p \ b	10	12	14	16	18	20
			Ni, $\lambda=44.4$ Å			
65	144	168	173	159	130	94.8
67	85.9	113	122	112	84.9	53.2
69	87.7	109	121	117	95.0	63.4
71	82.1	103	117	119	103	76.7
73	75.9	95.8	111	127	115	86.7
75	77.1	98.0	115	123	117	94.7
77	70.2	89.8	106	116	126	107
79	62.8	91.5	109	121	123	110

L_p \ b	4	6	8	10	12	14
			Ni, $\lambda=31.36$ Å			
65	30.6	60.3	85.8	92.6	78.3	50.4
67	18.4	36.0	53.6	64.1	57.8	33.1
69	17.0	33.5	54.8	67.3	62.5	39.3
71	17.2	33.9	51.5	64.9	65.0	45.2
73	15.7	31.2	47.8	61.4	65.1	50.2
75	14.2	28.3	48.6	63.2	69.2	57.4
77	14.3	28.6	44.4	58.4	66.1	59.6
79	12.7	25.5	39.8	52.9	69.1	65.5

L_p \ b	2	3	4	5	6	7
			Au, $\lambda=9.89$ Å			
65	6.7	11.8	14.2	12.6	8.5	5.0
67	3.9	6.9	8.8	8.0	4.8	2.3
69	3.6	6.5	8.5	8.3	5.4	2.8
71	3.7	6.6	8.8	9.0	6.4	3.3
73	3.4	6.1	8.3	8.9	6.8	3.7
75	3.1	5.6	7.7	8.5	7.1	4.1
77	3.1	5.7	7.9	9.0	7.9	4.9
79	2.8	5.1	7.2	8.3	7.7	5.2

L_p \ b	4	6	7	8	10	12
			Au, $\lambda=21.61$ Å			
65	19.9	29.6	31.3	30.9	25.5	18.3
67	10.8	15.2	15.7	15.0	10.8	6.7
69	10.1	14.5	16.7	16.2	12.3	7.3
71	10.3	15.1	16.1	15.9	12.6	7.9
73	9.6	14.2	15.2	15.3	12.7	8.2
75	8.7	13.1	14.2	16.2	13.9	9.4
77	8.9	13.5	14.8	15.2	13.5	9.6
79	8.0	12.2	13.5	14.0	12.8	10.7

L_p \ b	4	6	8	10	12	14
			Au, $\lambda=31.36$ A			
65	20.6	33.2	40.4	41.2	37.4	31.5
67	11.1	16.6	18.7	17.4	15.1	10.7
69	10.3	15.6	19.8	18.8	15.4	11.2
71	10.6	16.2	18.9	18.4	15.5	11.5
73	9.8	15.1	17.8	17.7	15.2	11.6
75	8.9	13.9	18.7	18.8	16.6	13.0
77	9.1	14.3	17.3	17.7	15.9	12.7
79	8.1	12.9	15.7	16.3	16.1	14.0

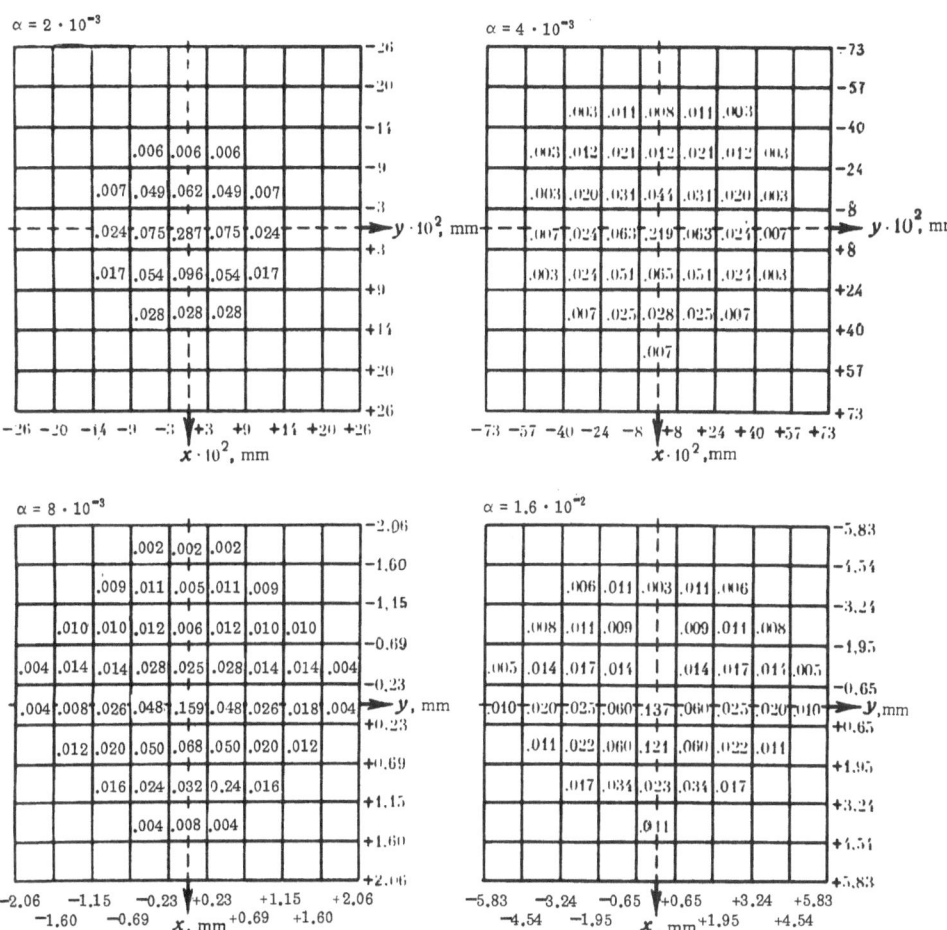

Fig. 17. Distribution of intensity in focal plane for objective (parabola + hyperbola) of Ni, b = 10 cm, L_p = 70 cm, F = 65 cm, λ = 31.36 Å.

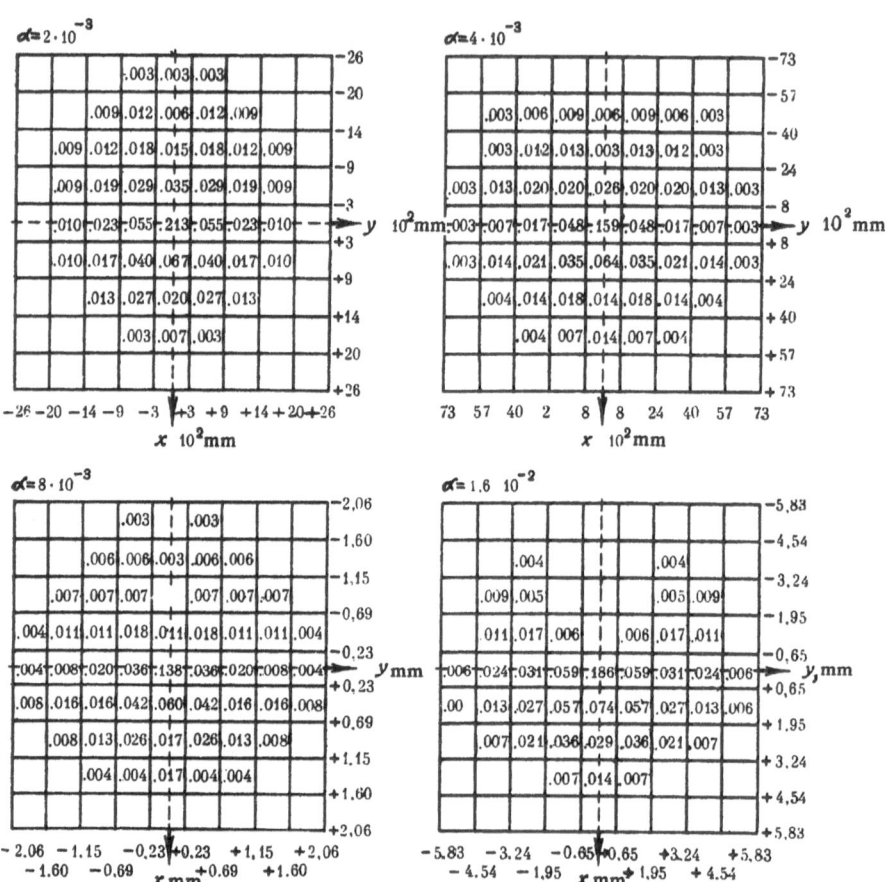

Fig. 18. Distribution of intensity in focal plane for objective (parabola + hyperbola) of Au, b = 10 cm, L_p = 6 cm, F = 65 cm, λ = 31.36 Å.

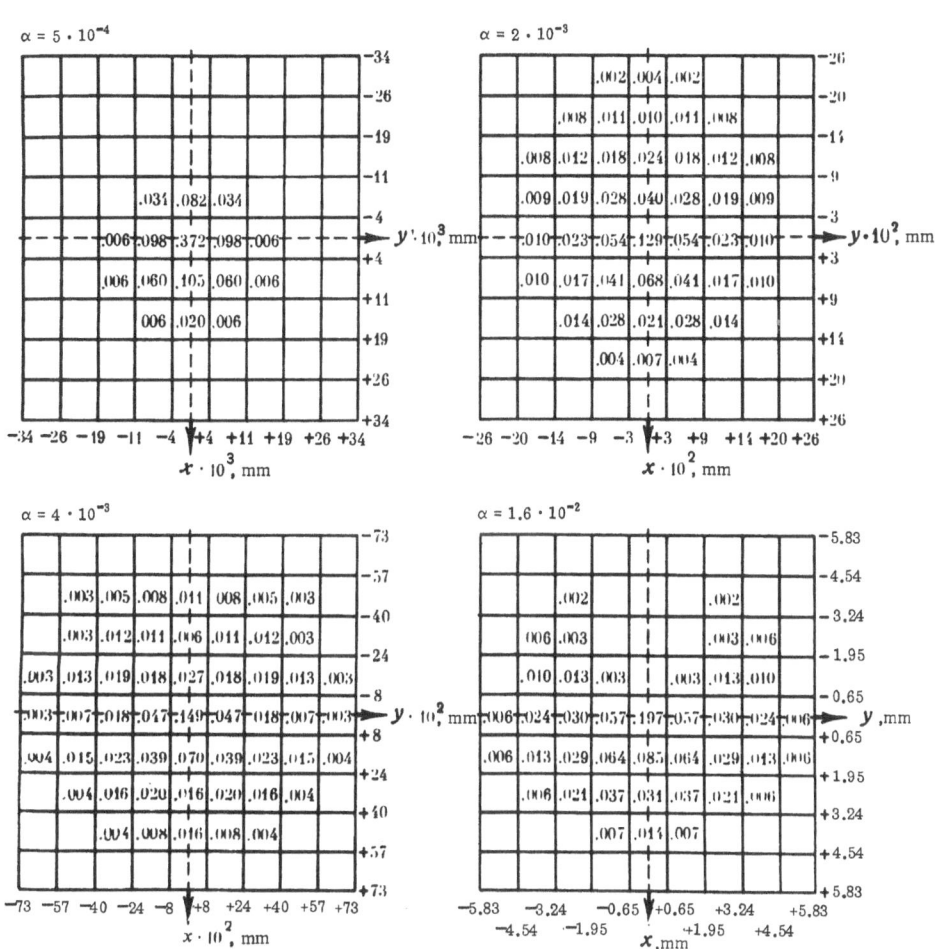

Fig. 19. Distribution of intensity in focal plane for objective (parabola + hyperbola) of Au, b = 6 cm, L_p = 70 cm, F = 65 cm, λ = 12.25 Å.

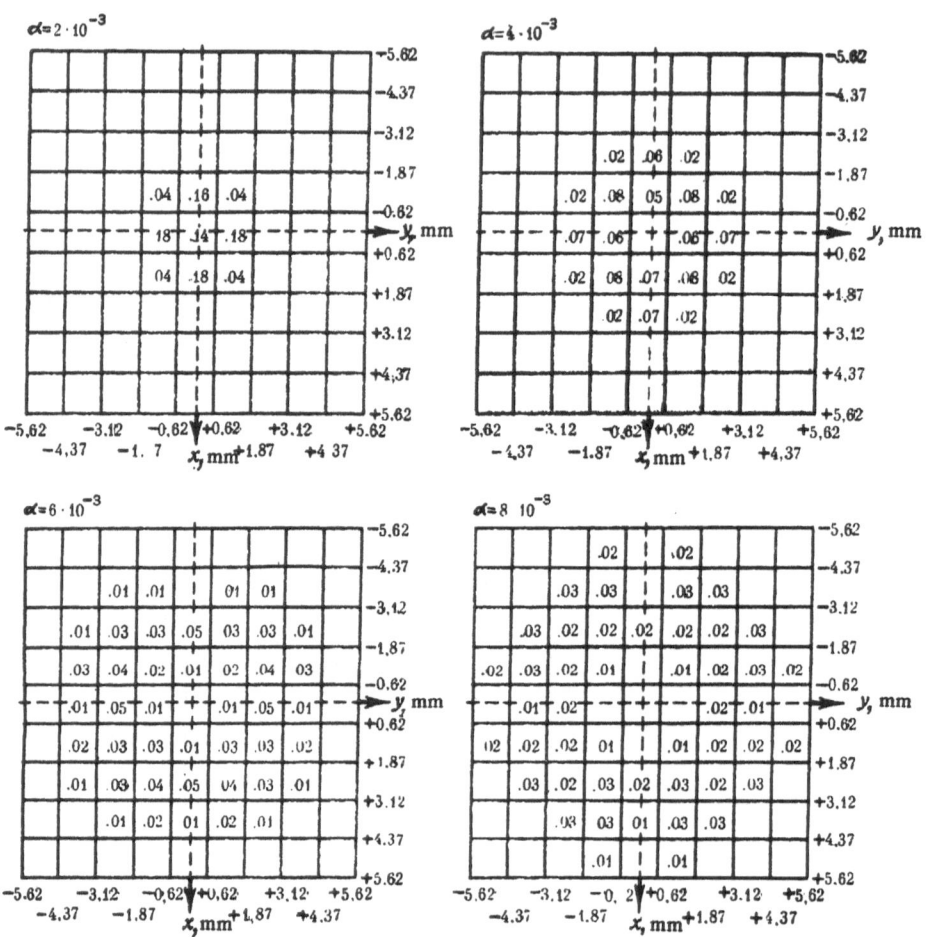

Fig. 20. Distribution of intensity in focal plane for paraboloid of Ni, b = 10 cm, L_p = F = 65 cm, λ = 31.36 Å.

aberrations can be judged from Fig. 16. The relative distribution of intensity in the image plane at a fixed angle α between the incident parallel beam with λ = 44.4 Å and the axis is shown in it.

The calculation was carried out for an objective of Ni with an entrance diameter 2b = 32 cm, a focal distance L_p = 70 cm for the paraboloid, and a focal distance F = 65 cm for the system. The angle α = is given in radians. The center of the frame bounding the field of view follows the center of the image, which is shifted from the main optical axis in the positive direction of the x axis according to the law $\alpha L_p/2$.

The law of variation in the frame dimensions $2^8 \alpha^{3/2}$, which corresponds approximately to the increase in aberrations, is established empirically.

From Fig. 16 it is easy to find the size at halfwidth for each of the angles presented.

The results of a calculation (for λ = 31.36 Å) of an objective of Ni with entrance diameter 2b = 20 cm, focal distance L_p = 70 cm for the paraboloid, and focal distance F = 65 cm for the system are presented in Fig. 17 as an illustration of the dependence of the aberrations on the wavelength (entrance diameter).

Analogous calculations for objectives of Au with entrance diameter 2b = 12 cm, L_p = 70 cm, and F = 65 cm for wavelengths of 31.36 and 12.25 Å are presented in Figs. 18 and 19, respectively.

The distribution of intensity for a paraboloid alone of Ni with an entrance diameter 2b = 20 cm, focal distance L_p = 65 cm, and at the wavelength of 31.36 Å is given in Fig. 20 for comparison.

5. DEMANDS ON ACCURACY OF FABRICATION

The development of x-ray astronomy is connected to a considerable extent with improvement in the quality of fabrication of the x-ray optics. Both the efficiency of the objective (the quality of the polishing strongly affects the reflection coefficient) and its resolving power depend on the quality of fabrication.

Let us dwell on the accuracies of fabrication which lead to limitations in the resolution of mirror systems. Deviation from the assigned dimensions in the observance of the shape of the surface does not it itself lead to marked distortions. The errors arising during fabrication can be divided arbitrarily into two groups: errors which change the dimensions without a change in the slope of the generatrix, and errors connected with undulation of the surface, which change the direction of the reflected ray. The first group of errors includes: the entrance and exit diameters not being coaxial, deviation in the dimensions of the diameters (ΔR), roundness — deviation from the shape of a circle. The limitations on this class of errors are usually rather mild.

Errors of the second group are angular and therefore directly related to the angular resolution of the objective. Their recalculation into linear deviations leads to more rigid demands on the accuracy of fabrication.

The demands on the accuracy of fabrication of an objective needed to obtain a resolution of ~1" are given in [6]. The allowable relative deviations in the fabrication of an objective with a resolution of ~1' are presented below.

We introduce the following designations: δx is the linear deviation in the image plane, y is the ordinate (the distance from the optical axis) of the reflecting surface, $y' = \tan\alpha \approx \alpha$ is the angle of inclination of the reflecting surface, and $\delta y'$ is the error in the angle. Then it is easy to obtain

$$\frac{\delta x}{y} = \left|\frac{\delta y}{y}\right| + \left|\frac{\delta y'}{y'}\right|. \tag{14}$$

The first term on the right side of Eq. (14) is the errors connected with the deviation in linear dimensions. Taking the angular resolution $\delta\alpha = (2/L_p)\delta x = 3 \cdot 10^{-4}$ (an angular minute), where L_p is the focal distance of the paraboloid, from Eq. (14) the maximum allowable total of relative deviations of the roundness, ellipticity, axial nonalignment, etc. types is

$$\frac{2}{L_p}\,\delta y = 3 \cdot 10^{-4}.$$

For an objective with $L_p = 70$ cm we obtain $\delta y \approx 0.1$ mm.

The second term of Eq. (14) corresponds to relative angular errors. For an angular resolution of $\delta\alpha = 3 \cdot 10^{-4}$ we have

$$3 \cdot 10^{-4} = \frac{2}{L_p}\,\delta x = \frac{2y}{L_p}\left(\frac{\delta y'}{y'}\right). \tag{15}$$

If the angular error is caused by an undulation in the surface of the type $\delta y = (h/2)\sin 2\pi\,(l/\lambda)$, where h is the height of the irregularities, λ is the period, and l is the current coordinate along the generatrix, then

$$[(\delta y_l')]_{\max} = \pi\frac{h}{\lambda}. \tag{16}$$

Substituting (16) into (15), for an objective with $(y/y')_{\max} = 2F = 2.65$ cm and $L_p = 70$ cm at a resolution of 1' we obtain

$$\frac{h}{\lambda} \leqslant \frac{\delta_2 L_p}{4\pi F} = 0.2 \cdot 10^{-4}$$

6. PROSPECTS OF APPLICATION

a) Paraboloid

The use of a paraboloid for the study of soft x radiation gives a number of advantages over the usual counter telescopes not only in the increase in the signal-to-noise ratio.

The focusing properties of the paraboloid allow a considerable decrease in the entrance window of the counter, which is especially important for the recording of radiation with E < 0.3 keV. For this energy region it is necessary to use very thin plastic films as the entrance window of the counter. The micropores which exist in them allow the leakage of gas from the volume of the counter and decrease the reliability of the operation of the entire system. Therefore the maximum decrease in the area of the entrance window is extremely desirable.

In addition, the application of the paraboloid allows the use of other types of detectors such as secondary multipliers, semiconductor detectors, channel multipliers, etc.

The principal results obtained up to now in reflecting x-ray astronomy are due to the use of a proportional flow counter as the detector. The main difficulties which exist here are connected with searches of new thin films which maintain a rather high pressure under conditions of prolonged operation in space and in addition possess good transmission in the region of

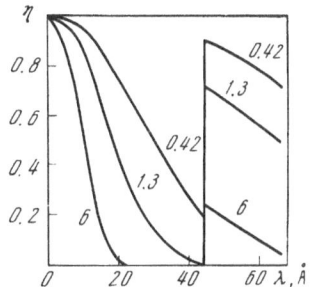

Fig. 21. Transmission of polypropylene films
of different thickness.

TABLE 3

| λ, Å | $S_{eff}\eta_{fi}$, cm² | | Optimum radius b, cm |
	thickness of counter window 6 μ	thickness of counter window 1.3 μ	
21.6	~0	18	6
31	~0	11	10
44.4	45	130	13
67	22	220	21

TABLE 4

| λ, Å | $S_{eff}\eta_{fi}$, cm² | | Optimum radius b, cm |
	window thickness 1.3 μ	window thickness 0.42 μ	
8.34	11	11.5	4
9.89	12.5	13.5	4
12.25	15	17.7	5
21.6	11.5	22	7
31	5	19.5	10

20-100 Å. The dependence of the transmission coefficient η on the wavelength for films of polypropylene 6, 1.3, and 0.42 μ thick is presented in Fig. 21. The total efficiency of the telescope was calculated on the basis of these data and the parameters of a paraboloid giving the maximum effective area at the given focus F = 65 cm (see Section 2). The results of the calculations for a nickel coating as a function of the wavelength are presented in Table 3. The optimum radius b of the entrance window of the paraboloid is given for each efficiency.

Similar results for a paraboloid with a gold coating are presented in Table 4.

The use of a semiconductor detector, such as silicon doped with lithium Si(Li), allows one to abandon the complicated gas system of a flow counter. But difficulties connected with the necessity of cooling the detector arise here. The efficiency of a detector of Si(Li) in the region of 8-50 Å is determined by the transmission of the surface layer of Si and of the filter which absorbs the visible light. The wavelength dependence of the transmission coefficient of

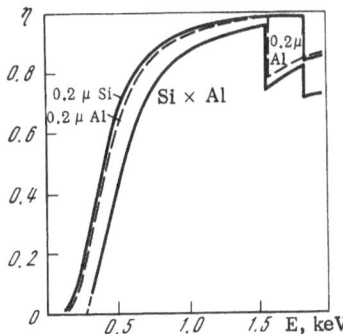

Fig. 22. Transmission of silicon and aluminum
and total efficiency of a semiconductor detector.

TABLE 5

λ, Å	$S_{eff}\eta_{s.c.d.}$, cm^2			
	parabola of Au	optimum value of b, cm	parabola of Ni	optimum value of b, cm
8	11.4	4	—	—
10	13	4	—	—
12	16.8	5	—	—
22	15.7	7	25	6
31	9	10	20	10
44	—	—	7.5	13
67	—	—	~0	21

0.2 μ Si and 0.2 μ Al and the total efficiency η of a semiconductor detector* are presented in Fig. 22. The maximum efficiency of 65-cm nickel and gold telescopes with a semiconductor for a number of wavelenths is presented in Table 5. The optimum radius b of the entrance pupil of the paraboloid is given for each efficiency.

Recently a lot of attention has been paid to the use of channelotrons as detectors of x radiation. Channel multipliers differ advantageously from counters and semiconductor detectors (s.c.d.) in small dimensions, the absence of an entrance window (it is true that a filter is needed for visible radiation), and the high signal level at the output.

b) Obtainment of an X-Ray Image

The next and most difficult step consists in obtaining an x-ray image of the source. The main problem here is undoubtedly connected with the fabrication of high-quality x-ray optics. The accuracies needed to obtain an image with a resolution of ~1' can be achieved on high-precision metal-working programmed lathes, but one-second resolution lies at the limits of the possibilities of modern optical technology.

Serious problems arise also in the development of x-ray pickups with high spatial resolution. For example, in using a paraboloid and hyperboloid with a length of 1 m from the entrance pupil to the focus of the system as the objective one must have a detector with a resolution of ~6 lines/mm to obtain a resolution of 1'. Such a spatial resolution is now possessed by electron-optical converters and microchannelotron disks.

APPENDIX

The equations are

$$p = -L_p + \sqrt{L_p^2 + b^2}, \tag{A.1}$$

$$z_p = \frac{b^2}{2p}. \tag{A.2}$$

The conditions for parabola and hyperbola to be confocal are

$$\frac{L_p - F}{2} = \sqrt{a^2 + c^2}, \tag{A.3}$$

$$z_h = F + \frac{L_p - F}{2} = \frac{F + L_p}{2}. \tag{A.4}$$

---

* These data pertain to a detector developed in the laboratory of physical electronics of the Institute of Nuclear Research (France). The authors are grateful to Dr. L. Koch for their presentation.

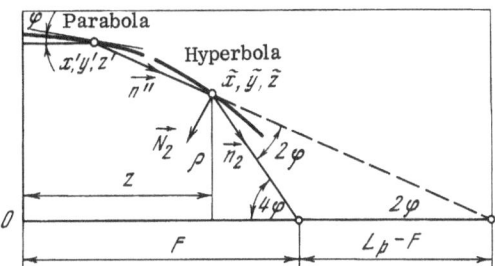

Fig. 23. Ray path for parabola + hyperbola objective.

The constants a and c are found from the condition of equality of angles for the central ray (passing through the center of the system; see Fig. 23):

We find a and c by substituting these equations into the equation for the hyperbola.

We designate

$$a = \cos \gamma \sqrt{a^2 + c^2}, \tag{A.4a}$$
$$c = \sin \gamma \sqrt{a^2 + c^2}, \tag{A.4b}$$
$$u = \sin^2 \gamma.$$

The equation for u is

$$\frac{u^2}{4} + u\,(1 + \cos 4\varphi) - \sin^2 4\varphi = 0,$$
$$u^2 + 8u - 64\varphi^2 = 0, \tag{A.5}$$

$$u = -4\,(-1 + \sqrt{1 + (2\varphi)^2}) \simeq 8\varphi^2,$$
$$u = \sin^2 \gamma \simeq \gamma^2 = 8\varphi^2, \quad \gamma = 2\sqrt{2}\,\varphi, \quad \varphi \simeq \frac{b}{2L_p}, \quad \gamma = \frac{\sqrt{2}\,b}{L_p}. \tag{A.6}$$

In the program $\gamma = \gamma_0 \varepsilon$, where γ_0 is determined from (A.6) and ε is varied: $0.8 \lessgtr \varepsilon \lessgtr 1.2$.

After finding a and c we can determine the length z_1 of the parabola:

$$\frac{(z - z_1)^2}{a^2} - \frac{p^2}{c^2} = 1, \quad \underbrace{c_1 z_1^2}_{a_1} - 2z_1 \underbrace{(c_1 z_h - c_2)}_{a_2} + \underbrace{c_1 z_h^2 - 2c_2 z_p - 1}_{a_3} = 0,$$
$$p^2 = 2p\,(z_p - z), \quad c_1 = 1/a^2, \quad c_2 = p/c^2, \quad c_2 \gg c_1, \tag{A.7}$$
$$z_1 = \frac{a_2 - \sqrt{a_2^2 - a_1 a_3}}{a_1}.$$

The coordinates of the point of intersection of a ray reflected from the parabola and hyperbola (see Fig. 23) are

$$\tilde{x} = x' + \frac{n_x''}{n_z''}\,(\tilde{z} - z'),$$
$$\tilde{y} = y' + \frac{n_y''}{n_z''}\,(\tilde{z} - z'), \quad \frac{n_x''}{n_z''} = \alpha, \quad \frac{n_y''}{n_z''} = \beta, \quad \tilde{z} - z' = f.$$

Substituting $\tilde{x}$, $\tilde{y}$, $\tilde{z}$ into the equation for the hyperbola, we obtain for f:

$$f = \tilde{z} - z' = \frac{-a_2 + \sqrt{a_2^2 - a_1 a_3}}{a_1},$$

$$a_1 = \frac{1}{a^2} - \frac{a^2 + \beta^2}{c^2}, \qquad a_2 = \frac{z' - z_h}{a^2} - \frac{x'a + y'\beta}{c^2},$$

$$a_3 = -\frac{x'^2 + y'^2}{c^2} + \frac{(z' - z_h)^2}{a^2} - 1.$$

The normal N_2 at the point $\tilde{x}$, $\tilde{y}$, $\tilde{z}$ is

$$N_2\left(\frac{m_1}{N}, \frac{m_2}{N}, -\frac{1}{N}\right),$$

$$m_1 = \frac{a^2 \tilde{x}}{c^2(\tilde{z} - z_h')}, \qquad m_2 = m_1 \frac{\tilde{y}}{\tilde{x}}, \qquad N = \sqrt{m_1^2 + m_2^2 + 1}.$$

The unit vector of the ray reflected from the hyperbola is

$$n_2 = 2(n''N_2)N_2 - n'' \quad \text{(reflection theorem)}.$$

We define $n''N_2 = g$. Then the projections of the vector n_2 on the coordinate axes are

$$n_{2x} = 2gN_{2x} - n''_x, \quad n_{2y} = 2gN_{2y} - n''_y, \quad n_{2z} = 2gN_{2z} - n''_z = -\frac{2g}{N} - n''_z.$$

LITERATURE CITED

1. D. J. Yentis, R. Novick, and P. Vanden Bout, Astrophys. J., 177, Part 1, 365-373, 375-386 (1972).
2. D. J. Yentis, J. R. P. Angel, D. Mitchell, R. Novick, and P. Vanden Bout, New Techniques in Space Astronomy (IAU Symposium 41, 1971), pp. 145-158.
3. I. L. Beigman, L. A. Vainshtein, Yu. P. Voinov, D. A. Goganov, N. I. Komyak, S. L. Mandel'shtam, I. P. Tindo, N. A. Shatskii, and A. I. Shurygin, Tr. FIAN, 73, 3 (1974).
4. H. Wolter, Ann. Phys., 10, 94 (1952).
5. H. Wolter, Ann. Phys., 10, 286 (1952).
6. R. Giacconi, W. P. Reidy, G. S. Vaiana, L. P. Van Speybroeck, and T. F. Zehnpfennig, Space Sci. Revs, 9, 3 (1969).
7. J. D. Mangus and J. H. Underwood, Appl. Opt., 8, 95 (1969).
8. G. S. Vaiana, W. P. Reidy, T. Zehnpfennig, L. P. Van Speybroeck, and R. Giacconi, Science, 161, 564 (1968).
9. L. P. Van Speybroeck, R. C. Chase, and T. F. Zehnpfennig, Appl. Opt., 10, 945 (1971).
10. O. A. Ershov, I. A. Brytov, and A. P. Lukirskii, Opt. i Spektr., 22, 127 (1967).
11. A. P. Lukirskii, E. P. Savinov, O. A. Ershov, and Yu. F. Shepelev, Opt. i Spektr., 16, 310 (1964).

EXTRA-ATMOSPHERIC STUDIES IN THE SUBMILLIMETER RANGE USING ON-BOARD TELESCOPES

A. E. Salomonovich and A. S. Khaikin

INTRODUCTION

This article is devoted to an account of a program of astronomical and aeronomical studies in the submillimeter wavelength range which can be carried out using on-board telescopes of moderate size and receivers presently developed. We will first examine some important problems of extra-atmospheric measurements and then turn to the current state of on-board receiving technology for the submillimeter range, dwelling mainly on accomplished projects.

Following this we will describe in more detail a program of observations and its resulting demands on the on-board submillimeter telescope with a main mirror of 1-1.5 m (the maximum attainable at present) and on the receivers for this telescope.

In a number of the following articles of this volume the main systems of the telescope are examined in detail: its optical system [1], receivers [2] and the cryogenic system required for their cooling [3], problems of filtering submillimeter radiation [4], and the construction of a model of an on-board telescope [5]. The results of the development of radiometers, intended for extra-atmospheric studies and to a large extent serving as prototypes and working models for the telescope, on which structural and technological problems arising during the creation of on-board submillimeter telescopes were studied, are presented in [6, 7].

1. ASTRONOMICAL AND AERONOMICAL PROBLEMS OF

EXTRA-ATMOSPHERIC SUBMILLIMETER STUDIES

In recent decades in addition to the classical region of visible light the armament of the sciences studying outer space and our planet has been supplemented by a number of new ranges not used earlier: the x-ray, gamma, and ultraviolet as well as infrared and radio regions of the electromagnetic spectrum.

On the entire scale of the electromagnetic spectrum there evidently remains at present only one region which has hardly been used yet for astronomical, aeronomical, and other studies. This is the region intermediate between the infrared and radio regions. This region, called the submillimeter or far infrared region, is located on the scale of the electromagnetic spectrum between the wavelengths of 100 μ (0.1 mm) and 1000 μ (1 mm). Sometimes the boundaries of the region are moved to 50 μ and 2 mm. These changes are usually determined by means of generation and reception of the radiation. Thus, the submillimeter region passes into the infrared region from the short-wave side and into the region of millimeter waves from the long-wave side.

The reasons why the submillimeter region has entered the arsenal of space research later than the others are of twofold origin:

a) the technology of the generation, reception, channelization, and analysis of submillimeter radiation is very complex and has developed more slowly than in the other ranges (this will be discussed below);

b) the conditions for the propagation of electromagnetic waves of this range on earth are extremely unfavorable. The radiation arriving from space hardly reaches the earth's surface at all (at sea level). For the use of submillimeter radiation it is necessary to go beyond the limits of the earth's atmosphere, which became possible only after the appearance of rocket-space methods.

The unfavorable conditions for the propagation of submillimeter waves in the earth's atmosphere have retarded the application of this range, which in turn has retarded the development of methods for their generation and detection.

At the start of the sixties, thanks to the successes of space technology and the physics and technology of semiconductors, electronics, and cryogenics, it became possible to enter upon the utilization of the submillimeter wave range for astronomy, aeronomy, and related fields. Both in the USSR and abroad the technology of submillimeter studies has developed rapidly in the last decade. This allows one to count on the fact that in the next few years a considerable advance will be made in extra-atmospheric submillimeter research.

Let us enumerate some of the main scientific problems for the solution of which observations in the submillimeter wave range are necessary or very desirable [8].

a) Cosmology. Study of the Characteristics of the Initial State of Matter

The theory of the expanding Universe has predicted the possibility of the existence of isotropic electromagnetic thermal radiation corresponding to the radiation of a black body which is at a temperature of about $3°K$. At an early stage of expansion of the Universe the density of the powerful radiation in the compressed hot plasma exceeded the density of the matter by many times. Because of subsequent expansion the energy of the radiation quanta decreased, although their number was conserved. At present the density of this isotropic radiation should exceed the density of the radiation of all other sources in the wave range where the radiation is maximal. It is easy to show that this maximum is located near 1 mm, while the Wien branch of the spectral curve lies in the submillimeter region. The so-called relict background radiation discovered by ground observers in the radio range in 1965 is an important subject for submillimeter studies. From the nature of the spectrum of this radiation, mainly in the Wein region, one can judge the separate stages of the evolution of the Universe [8].

Measurements of the relict background radiation in the submillimeter range are extremely complicated. They require lifting the apparatus beyond the limits of the earth's atmosphere and deep cooling not only of the receivers (see below) but also of the entrance optical system in the avoidance of parasitic irradiations. The results of the first attempt at extra-atmospheric measurements of the background, undertaken starting in 1968 in the USA using geophysical rockets of the Aerobee type [9] and since 1970 also on the solid-fuel Terrier-Sandhook rocket [10], show how difficult these experiments are. The effective temperatures of the black-body radiation estimated from these measurements were in considerable disagreement with the data of ground measurements [8] and with the results of a measurement made in 1969 using a radiometer in the submillimeter range raised on a balloon to an altitude of 40 km [11]. Whereas the aggregate of ground measurements in the radio range agree with the hypothesis of the presence of an isotropic background corresponding to a black-body temperature of 2.7°K, the rocket measurements made in 1968, 1969, and 1971 in a wide band from 0.4 to 1.3 mm [9] led to a flux of background radiation exceeding the expected flux at this temperature $(4 \cdot 10^{-11}$ $W/cm^2 \cdot sr)$ more than 30-fold. A measurement made by the same group [9] in 1972 led to a smaller flux $(12 \pm 20 \cdot 10^{-11} W/cm^2 \cdot sr)$, although also exceeding threefold the expected flux at $T = 2.7$°K. The results of balloon measurements [11] in the same range can be interpreted as the superposition of background thermal radiation at a temperature of about 2.7°K and intense monochromatic radiation at a frequency of 11-12 cm^{-1} (a wavelength of 800-900 μ). Without this additional radiation the background recorded in [11] would correspond to black-body radiation at ~ 6°K.

A rocket measurement of the background [10] in the range of 6-0.8 mm leads to "black" radiation at a temperature of $3.1^{+0.5}_{-2.0}$°K and the absence of the monochromatic radiation needed for the interpretation of the results of [11]. However, the very accuracy of the measurements in [10] and the fact that these measurements cover only the long-wave section of the spectrum reduce the reliability of this result.

The principal conclusion arrived at by all the authors consists in the necessity of extending the measurements of the background relict radiation. Measurements of the radiation spectrum at wavelengths $\lambda \lesssim 1$ mm are especially important here.

b) Studies of the State and Chemical and

Isotopic Composition of the Intergalactic

and Interstellar Medium

As mentioned above, the maximum in the spectral density of the thermal radiation corresponding to temperatures on the order of 10-20°K falls in the submillimeter region. Therefore this region is appropriate for the study of very cold regions of the Galaxy. The measurement of the intensity distribution of submillimeter radiation is of interest for the determination of regions of gravitational condensation of matter and of dust formations. At wavelengths of about 200 μ in particular one can expect a maximum in the intrinsic radiation of interstellar dust (if its temperature, as may be assumed, is close to 20°K).

Spectral measurements are very promising in the submillimeter region since many resonance lines of rotational transitions of molecules of hydrogen, oxygen, nitrous oxide, cyanogen, hydroxyl, water, and many others are located in this region of the spectrum. Here, as in the millimeter range, one can expect maser effects. A number of lines of excited atomic hydrogen and other elements, corresponding to transitions between nearby energy levels with large quantum numbers (recombination lines), are located in this range. In 1968 such a line (H56α) was discovered in the Omega nebula in the millimeter range [12]. Measurements of the intensity, frequency, and width of lines of recombination radiation represent an effective means of studying the distribution of concentration and the electron temperature as well as the velocities of movement of regions of ionized interstellar gas.

The spectral density of radiation in a line of the submillimeter range, as calculations show, should considerably exceed the density of thermal radiation of the galactic continuum in the same section of the spectrum, although the brightness temperature of recombination lines is lower than at centimeter wavelengths where these lines were first detected [13].

c) Study of Discrete Sources — Quasars, Seyfert Galaxies, Infrared Stars

One of the most promising problems of submillimeter extra-atmospheric research is the detailed study of the spectra of the extremely large number of unusual cosmic objects discovered in recent times. Such objects primarily include quasars, Seyfert galaxies, and the so-called "infrared stars" which are invisible in the optical range and were recently discovered by measurements in the IR-range.

A noteworthy property of these and some other objects is their unusual spectra: One can assume the presence of a maximum in the spectral density of the radiation right in the submillimeter range, just as infrared stars have a maximum in the IR-region from 3 to 20 μ.

Such a nonmonotonic spectrum can be assumed in particular for the well-known source of radio radiation Cygnus-A. The data on the flux of its radiation obtained recently by ground measurements in the region between 3 and 1 mm [14-18] are rather contradictory. According to some of them [15, 16] a sharp increase in the spectral density of the flux is observed at wavelengths near 1 mm; according to other data [17, 18] there is evidently no such increase and the flux density declines more or less monotonically from long to short wavelengths up to the IR-region. Nevertheless, because of the large errors of ground submillimeter measurements the question cannot be considered as solved and requires detailed spectral measurements at wavelengths near 1 mm and shorter. The state and the problems of measurements of discrete sources will be analyzed in more detail below, in Section 2, in connection with the development of a program of experiments with an on-board submillimeter telescope.

d) Studies of Celestial Bodies of the Solar System

Studies of the submillimeter radiation of the sun and planets of our solar system are of special interest. Depending on the physical processes responsible for the electromagnetic radiation of a celestial body the radiation at different wavelengths comes from different layers of the source. By conducting comparative measurements of the radiation intensity in different sections of the spectrum we can obtain information on the depth distribution of the material parameters of the body.

By measuring the intensity of the sun's submillimeter radiation we can obtain information on the deepest (closest to the photosphere) layers of its atmosphere, since the layers lying above them are transparent to submillimeter waves. The absence until recently of reliable information on the sun's brightness temperature at wavelengths shorter than 0.8 mm and down to 0.02 mm has prevented the construction of a valid model of its lower chromosphere. Even the recent aerostatic measurements (see below) do not yet give reliable data for such construction [19].

Measurements of the moon's brightness temperature at submillimeter wavelengths provide information on the physical conditions in its subsurface layers lying closest to the surface.

Studies of the distribution of brightness temperature over the disks of the sun and moon using instruments with sufficiently narrow-beam optics allow one to obtain information on nonuniformities, particularly on foci of solar activity connected with facular fields and flocculi. The first measurements of this kind, conducted under ground conditions [8] in the submillimeter range on radiotelescopes of the RT-22 type, showed the great possibilities of submillimeter measurements and at the same time their limitation by the conditions of propagation of submillimeter waves in the earth's atmosphere. The ascent to those altitudes where the effect of the absorption and self-radiation of the atmosphere are made negligibly small is necessary for fully adequate measurements at wavelengths at least shorter than 800 μ.

Spectral measurements of the sun's radiation in the submillimeter range are of great interest. Theoretical studies [20] show that the detection of recombination lines produced by the phenomenon of so-called two-electron recombination can serve as a means of determining the abundances of highly ionized atoms of heavy elements in the solar corona.

Planetary studies, in addition to the determination of the temperature conditions of planetary atmospheres at the levels where submillimeter radiation is efficiently absorbed, provide information on the chemical composition of the upper layers of planetary atmospheres: A multitude of bands of rotational transitions of the molecules of gases whose presence in the atmospheres of planets is assumed or established is located in the millimeter and submillimeter ranges.

e) Aeronomical Submillimeter Studies

As was noted above, the earth's atmosphere represents a medium which is extremely unfavorable for the observation of cosmic sources in the wavelength range shorter than 1 mm. Spectra of the absorption of water vapor of the earth's atmosphere calculated at sea level [25] are presented in Fig. 1. The density and temperature fall off with altitude, and width and intensity of the absorption lines decrease. Absorption spectra of the earth's atmosphere in the submillimeter range are presented in Fig. 2. The calculations were made for altitudes of 0.5 (curve 1), 3.5 (2), and 30 km (3) above sea level [26].

Besides the absorption of cosmic radiation the atmosphere contributes its own intrinsic background radiation. Fluctuations in the thermodynamic parameters (density, temperature) and in the coefficients of absorption and refraction which depend on them as well as the presence of turbulent movements in the atmosphere lead to fluctuations in the brightness temperature of the self-radiation. Liquid drops and solid particles suspended in the atmosphere — clouds and haze, as well as precipitation — hail, rain, and snow — introduce even greater variety. It is just because of these circumstances that extra-atmospheric studies of cosmic sources are so necessary.

On the other hand, such a connection between the parameters of the atmosphere and the intensity (and polarization) of the submillimeter radiation makes this range very promising for the study of the earth's atmosphere itself, at least its upper layers, the stratosphere.

Profiles of the brightness temperature made at a sufficient altitude in the vertical plane allow one to obtain information on the distribution of certain parameters of the atmosphere, while azimuthal "surveys" allow one to judge the heterogeneity of the atmosphere. Surveys of the intensity distribution of submillimeter radiation carried out along an orbital trajectory with the optical axis of the radiometer aimed at the nadir are especially interesting. The dependence of the brightness temperature on the longitude and latitude of the observation site permits one to estimate the variations in the parameters of the effectively radiating layer with the geographical coordinates and the differences between equatorial and polar and between sea and land regions. Such observations conducted in several relatively narrow spectral intervals can give material on the abundance of one or another component of the atmosphere in different regions. The first vertical profiles in a wide band in the submillimeter range (0.5-2 mm) from an altitude of 35 km were conducted at the P. N. Lebedev Institute of Physics of the Academy of Sciences of the USSR (IPAS) in 1968 [21]. Measurements of the spectra of the absorption of solar radiation by the atmosphere were conducted on an airplane from an altitude of 12 km in the region of 600-1500 μ [29] and from an altitude of 4.3 km at the high-mountain Mauna Kea station (Honolulu) in the region of 10-12 cm^{-1} (800-1000 μ). The latter observations are inter-

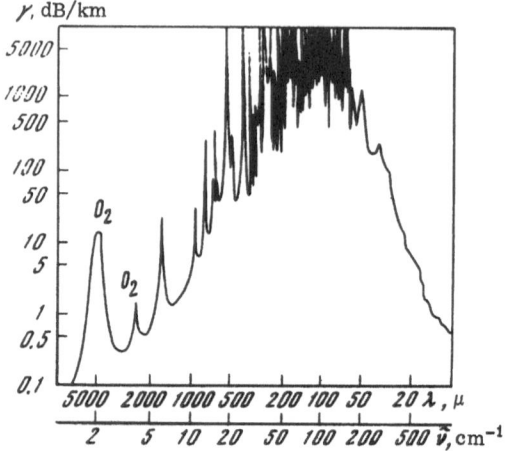

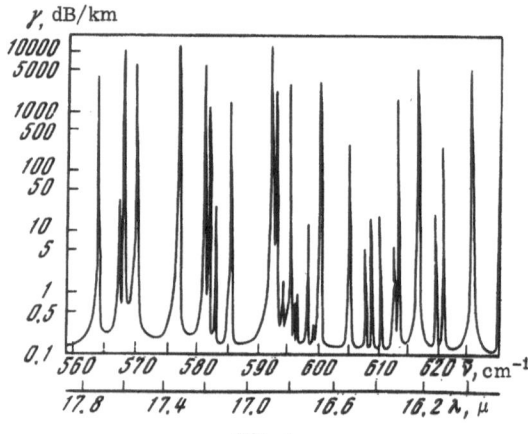

Fig. 1

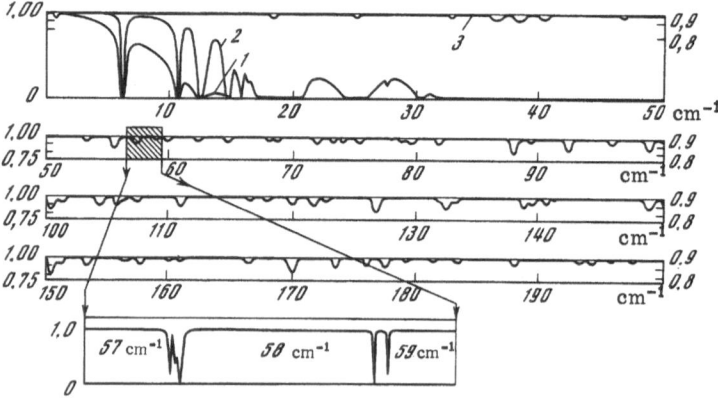

Fig. 2

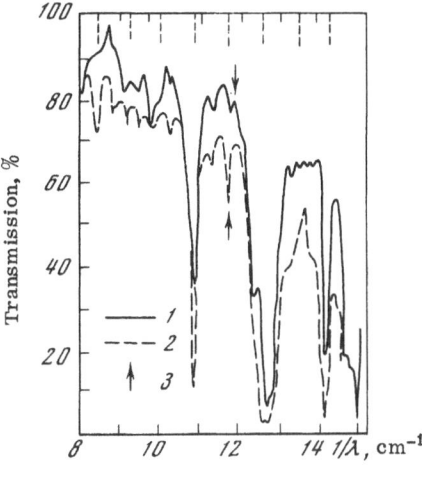

Fig. 3

esting in that on the spectrogram obtained in them a singularity is detected (Fig. 3) at $\nu = 11.7$ cm^{-1}, identified with the transition $J = 14 \to 13$ of the N_2O molecule. Notation in the figure: 1) radiation of sky; 2) absorption of sun's radiation; 3) N_2O line at 11.7 cm^{-1}. The nitrous oxide abundance (6.7 cm·atm) which follows from an analysis of the intensity of this line proved to be considerably greater than the expected abundance (0.13 cm·atm). While demonstrating the possibility of studying the components of the atmosphere in the submillimeter range, this result in addition may shed light on the contradictory data obtained during rocket and balloon measurements of the background radiation from space, mentioned in Part a.

Studies of the earth's atmosphere from satellites and geophysical rockets in different sections of the submillimeter range in conjunction with analogous IR and radio observations, such as those conducted in the USSR on the satellite Kosmos-243 [24], are in the most elementary state — they represent a new effective means of studying the composition and dynamics of the earth's atmosphere. In particular, spectral measurements near the absorption lines of hydrogen vapor, oxygen, ozone, and other components excite great interest.

Let us now examine the state of the technology suitable for extra-atmospheric submillimeter measurements, i.e., adapted for mounting on board aircraft, balloons, and spacecraft.

2. ON-BOARD RECEIVER TECHNOLOGY

FOR THE SUBMILLIMETER RANGE

The current state of laboratory technology for the generation, reception, and spectral analysis of submillimeter radiation is reflected with sufficient completeness in the series of reviews [8, 27, 28]. Radiometers based on detectors and superheterodyne superwide-band receivers with crystal mixers at the input have been developed and used for the long-wave part of the range (down to $\lambda = 0.5$ mm). From the other side methods of reception developed for the optical and near infrared ranges are used in the submillimeter range: optico-acoustic transducers, bolometers of various types, including superconducting bolometers cooled to the temperature of liquid helium. Bolometers and receivers of the photoresistance type containing semiconducting elements of germanium doped with different additives and of gallium arsenide

TABLE 1

Type of receiver	Working temperature, °K	Minimum detected power ΔP at $\tau = 1$ sec, W	Detector time constant τ, sec	$\overline{(\delta T^2)^{1/2}}$ ($\lambda = 0.5$ mm)	Frequency band Δf, MHz
Superheterodyne based on a crystal detector	300	$1.1 \cdot 10^{-13}$	10^{-9}	130	60
The same	300	$1.7 \cdot 10^{-13}$	10^{-9}	10	1200
Optico-acoustic transducer	300	$3 \cdot 10^{-10}$	0.015	$4.5 \cdot 10^{-1}$	$6 \cdot 10^4$
Carbon bolometer	2.1	$1 \cdot 10^{-11}$	0.01	$1.5 \cdot 10^{-2}$	$6 \cdot 10^4$
Germanium bolometer	2.15	$5 \cdot 10^{-13}$	$4 \cdot 10^{-4}$	$7.5 \cdot 10^{-4}$	$6 \cdot 10^4$
Superconducting bolometer	3.7	$3 \cdot 10^{-12}$	1.25	$4.5 \cdot 10^{-3}$	$6 \cdot 10^4$
Receiver of antimonous indium band-widened with a magnetic field	1.5	$1 \cdot 10^{-11}$	$2 \cdot 10^{-7}$	$1.5 \cdot 10^{-2}$	$6 \cdot 10^4$
Receiver of antimonous indium without a magnetic field	4.0	$4 \cdot 10^{-12}$	$3 \cdot 10^{-7}$	$1.5 \cdot 10^{-3}$	$6 \cdot 10^4$

and indium antimonide are most promising from the point of view of maximum sensitivity and low inertia in the submillimeter range.

The parameters of the main receivers presently used are presented in Table 1.

The use of gas lasers in the submillimeter range and tunnel-effect transducers in contact with superconductors (the Josephson effect) opens up prospects for the creation of superheterodyne submillimeter receivers with a very low noise level. However, only exploratory work is presently being conducted in these directions (see [29], for example).

Some of the necessary elements for the channelization, filtration, and spectral analysis of submillimeter radiation have been developed, particularly instruments of the Michelson and Fabry-Perot interferometer types.

It must be emphasized, however, that until recently there have been almost no basic instruments and elements of the receiver-amplifier and spectral technology in the submillimeter range suitable for conducting experiments under extra-atmospheric conditions, i.e., elements conventionally called on-board elements. Thus, although there have been certain prototypes of these instruments under laboratory conditions it does not seem possible, as a rule, to use them under the real conditions of extra-atmospheric measurements. In connection with this a number of developments have been carried out in the spectroscopy laboratory of the IPAS jointly with the cryogenics division of the IPAS which have made it possible to create flight mock-ups and models of submillimeter-range radiometers suitable for use on high balloons, geophysical rockets, and automatic orbital satellites [6, 21, 30]. Analogous developments have also been carried out abroad (USA, England, France) in 1967-1971 [31-39]. Below we shall briefly examine the main types of on-board submillimeter radiometers and telescopes.

a) Aircraft Radiometers and Telescopes

A submillimeter-range Fourier spectrometer for use on an airplane has been developed at the National Physics Laboratory (England) [22]. The region of the spectrum is 15-65 cm^{-1} (600-1500 μ). An optico-acoustic transducer is used as the receiver. The diameter of the objective (a spherical mirror) is 43 mm. The amplitude modulator is placed at the focus of the mirror. Flights have been carried out at an altitude of about 12 km. The spectrum of the absorption of solar radiation in the atmosphere has been measured.

A submillimeter Cassegrain telescope with a primary mirror 30 cm in diameter has been mounted on the NASA jet Lyra (USA) [32]. Its field of view is 13' in elevation and 10' in azimuth. Guidance is accomplished visually with an accuracy no worse than 6'. Diaphragm modulation with a frequency of 80 Hz is used (the axis oscillates through 14' in azimuth). A cooled germanium bolometer at a temperature of 1.8°K serves as the receiver. Observations in the band of 40-300 μ were conducted at an altitude of about 15 km. Changeable filters with cutoffs at 40 and 50 μ served to determine the transmission function of the system. The telescope is sensitive only to gradients of radiation and was used for observations of the center of the Galaxy.

The creation of a 36-inch (92 cm) aircraft IR-telescope is reported in a report of the NASA Ames Research Center [33] and detailed information is provided on its mechanical, optical, and other parameters. It is indicated in particular that the telescope is capable of operating in the wavelength range from 1 μ to 1 mm. The field of view is 14' and the vignetting of the area of the main mirror is 8%. The main parabolic and secondary hyperbolic mirrors are constructed of an alloy with a zero thermal expansion coefficient at 220°K. Diagram modulation is accomplished by the secondary mirror. The observations are conducted through a hatch which is open in flight. The accuracy of stabilization of the telescope axis during a 30-minute session is no worse than 6'. The flight program calls for ascent to an altitude of 13.7 km.

b) Balloon Radiometers and Telescopes

Reports on the following balloon radiometers (telescopes) used in the submillimeter range have been published up to now.

Following the initial attempts at observations at an altitude of 30 km of discrete sources using a balloon radiometer with a germanium bolometer cooled to 1.8°K (in the 300-450 μ band) [34], a group at the NASA Goddard Institute for Space Studies has developed and used balloon submillimeter telescopes with mirrors up to 30 cm in diameter [35, 36]. In the latest modification of the balloon telescope with a mirror 30 cm in diameter the germanium bolometer is located at the Newtonian focus. The cooling to 1.8°K is accomplished through the reduced pressure (~ 10 torr) which exists at the flight altitude (29 km). A system of cooled and "warm" filters provides a transmission band of $\Delta\lambda = 50 \mu$ with a transmission maximum (20%) at $\lambda = 100 \mu$. The width of the field of view (at the 0.5 level) is 12'. Modulation is accomplished through oscillations of the Newtonian mirror with a frequency of 20 Hz, displacing the field through 18'. The parasitic signal is compensated for by the electronic amplifier circuit. The gondola of the stratosphere balloon is stabilized in position angle by the gravitational field and in azimuth by the earth's magnetic field. The accuracy in guidance and stabilization was 6'. The telescope axis could be aimed in a fixed direction or placed in a scanning mode with a rate of 0.3 deg/sec by commands from earth. The scanning region was selected from ± 1 to ±10°. The combination of scanning and the diurnal rotation of the celestial sphere provided a relatively narrow region of overlapping of the scans (a source appeared on three to four successive passes).

The balloon submillimeter Cassegrain telescope of the Meudon Observatory (France) [37] with a diameter of 40 cm and with an optico-acoustic receiver is used for spectral studies of the sun's radiation. The composition of the telescope includes a Fourier spectrometer with a resolution of 0.3 cm^{-1} in the range of 50-2300 μ. The gondola of the statosphere balloon, rising to an altitude of 30 km, is stabilized by the means indicated above and then by solar or star pickups with an accuracy of 10-20".

For the measurement of the sun's brightness temperature in the region of 120-400 μ a group of Swiss scientists [19] has developed and raised on a balloon similar to the one described above a lamellar interferometer with a grating 10 × 10 cm in size, providing a resolu-

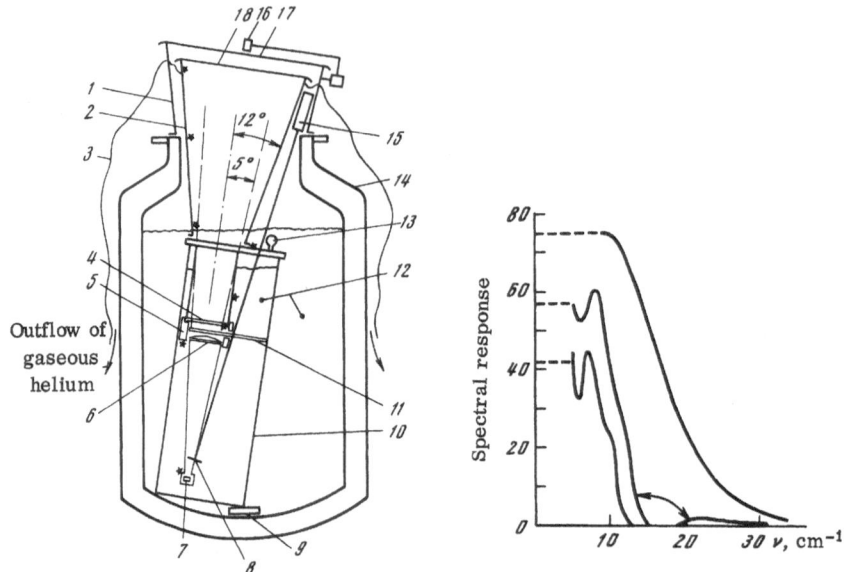

Fig. 4. Sketch of construction and spectral characteristics of radiometer. 1) Outer cone; 2) inner cone; 3) Mylar jacket; 4) cold window; 5) absorber; 6) Teflon lens; 7) In–Sb detector; 8) modulator; 9) heater; 10) copper vessel; 11) disk of interchangeable filters; 12) liquid helium 1.5°K; 13) safety valve; 14) Dewar vessel of stainless steel with a capacity of 25 liters; 15) modulator motor; 16) movable calibrator; 17) removable Mylar cover; 18) removable Mylar membrane. * Thermometer

tion of ~0.4 cm^{-1}. The interferometer was placed at the Newtonian focus of a telescope having a diameter of 20 cm. The sun's short-wave radiation was excluded by filters and by scattering on roughnesses of the deflecting mirror. The information was transmitted to the earth through a telemetry channel.

The balloon submillimeter radiometer of the P. N. Lebedev Institute of Physics [21] mentioned above, intended for measurements of the radiation of the atmosphere in the 500-2000 μ band, has a main mirror 120 mm in diameter. One of the plane tilting mirrors provides beam scanning with a width of 5° in the vertical plane according to a fixed program with periodic calibration against a standard black body. The indium antimonide receiver is cooled to 4.2°K in an on-board helium cryostat. The amplitude modulation necessary for atmospheric measurements is accomplished by a shutter located between the plane mirrors. The information is recorded on an on-board recorder.

A balloon submillimeter radiometer intended for the measurement of the intensity of the background radiation of space has been developed at the Massachusetts Institute of Technology (USA) [11]. A special feature of this radiometer is the fact that the entire instrument, including its optical section, is placed in a helium cryostat with a wide neck, which is necessary to prevent the strong "exposure" of the receivers by the radiation of the warm parts of the radiometer. A sketch of the radiometer construction is presented in Fig. 4. An indium antimonide detector is used as the receiver. Three filters introduced alternately into the channel cut off the short-wave radiation at different limiting frequencies. This allows one to judge the spectral distribution of the radiation flux density. A polyethylene film provides the maintenance of the necessary excess pressure of the evaporating helium. The film with a thickness of 10^{-2} mm is stretched in the form of a membrance in the field of view of the radiometer (~12°). In accordance with the program the cryostat can be inclined to limiting angles of 5-27° from the zenith.

c) Rocket Radiometers and Telescopes

For a number of problems, particularly for measurements of the relict background, and for studies of many weak discrete sources especially in the short-wave part of the submillimeter range, ascent to altitudes of 15-40 km proves to be insufficient. Radiometers and telescopes for geophysical rockets are developed as the next step. Among the instruments on which there is information we note the following.

In 1968 a submillimeter telescope was launched on an Aerobee rocket [9] to an altitude of 170 km by a group at Cornell University (USA). Like the balloon telescope of [11], the 170-mm reflecting telescope was placed in a cryostat. The cryostat cover cooled by helium was jettisoned with a powder cartridge after ascent to an altitude of 135 km. An indium antimonide receiver (400-1300 μ band) was located along with IR-detectors in the receiver chamber of the telescope placed at its focus. The telescope field of view was 5°. Amplitude modulation with a frequency of 150 Hz was accomplished using a tuning fork modulator. The telescope construction was improved in further experiments by this group, and in particular detectors of gallium arsenide (200-450 μ region) and of germanium doped with gallium (70-130 μ) were introduced in addition. Additional measures were taken to exclude the entrance of scattered radiation of the earth and of the instrument. All telescope calibrations were carried out in the laboratory before the launch.

A three-channel submillimeter radiometer mounted on the Terrier-Sandhook rocket was developed by a group at the Los Alamos Laboratory of the University of California (USA) for measurements of the background radiation of space [10]. The outstanding feature of this radiometer (Fig. 5) is the absence of reflecting optics and interchangeable filters. In their place three conical light pipes were placed in the volume of the cryostat, each of which contained tuning fork modulators with a frequency of 20 Hz and filters in three regions of the spectrum (with cutoffs at 1400, 700 and 100 μ on the short-wave side; the long-wave cutoff at λ = 6 mm in all channels is due to the confinement of the light pipe). Identical germanium bolometers served as the receivers. Cooling to 1.6°K was provided by automatic maintenance of reduced pressure in flight. Supplementary cryostats in which the temperature was kept at 4.2°K served to retain the cold. The upper cryostat was jettisoned together with the cover cone after reaching an altitude of 120 km. The field of view in each channel did not exceed 20° (at the 1% level). Calibration in flight was conducted with respect to the radiation of heated sections of the light pipes, while calibration under laboratory conditions was conducted with a cooled black body.

A submillimeter spectroradiometer (SMS) for measurements of the spectrum of the sun's radiation on the geophysical rocket Vertical was developed at the P. N. Lebedev Institute of Physics of the Academy of Sciences of the USSR in 1969-1970 and tested in 1971 [6]. The spectroradiometer consists of an on-board Cassegrain telescope (diameter of main mirror 160 mm) near the focus of which is placed a Michaelson interferometer (maximum path difference 20 mm). A cooled indium antimonide detector serves as the receiver. A control system using silicon differential solar pickups whose error signals control the movement of the mirror about both axes through a servomotor system is used for automatic tracking of the solar disk with an accuracy of ±8'. The interferometer and the cryostat containing the radiation receiver remain stationary in this case. The spectroradiometer field of view is 1° and the time for taking an

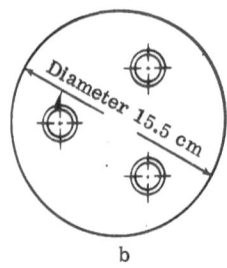

b

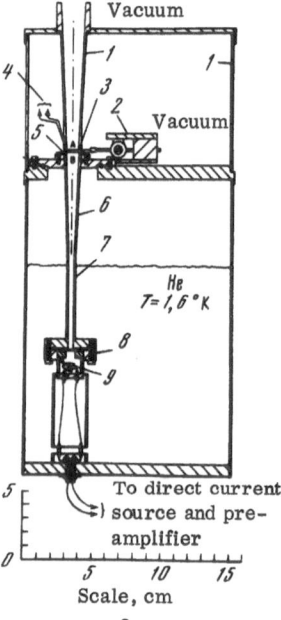

Fig. 5. Three-channel submillimeter radiometer.
a) Longitudinal cross section of radiometer: 1)
upper cone; 2) modulator; 3) heated section of
cone; 4) to heater current source; 5) filter;   6)
lower cone; 7) cylindrical section; 8) bolometer
lead-outs; b) transverse cross section.

To direct current
source and pre-
amplifier

Scale, cm

a

interferogram is about 1.5 min. The cryostat was charged 6 h before the flight. The spectroradiometer of the SMS is described in more detail in [6].

d)   Satellite Submillimeter Radiometers

and Telescopes

In the transition from aircraft and balloon radiometers to rocket and satellite instruments
the requirements which must be satisfied by the circuits and construction of the instruments
become more complicated. In fact, for balloon radiometers the main requirement is the automatic functioning (including aiming) under conditions of a relatively low outside temperature
and a gradually decreasing atmospheric pressure (down to 1 mm Hg).

Rocket radiometers must in addition withstand the g-forces produced on the active section of the flight and operate for a short time under weightless conditions on the passive section

Fig. 6. Model of satellite radiometer.

of the flight of the geophysical rocket. It is also desirable to preserve the instrument or its parts upon recovery.

In the development of radiometers and telescopes for orbital satellites we encounter additional requirements on the capacity for operation under conditions of prolonged weightlessness, a very deep external vacuum, and periodic oscillations in the external temperature.

We know of no results of developments of submillimeter radiometers and telescopes for orbital satellites which have been conducted abroad.

The development of the first laboratory and flight models of a satellite submillimeter radiometer with an indium antimonide receiver cooled to the temperature of liquid helium was carried out at the P. N. Lebedev Institute of Physics in 1969-1971. These developments led to the creation of models of the Obzor radiometer, with the help of which the cryogenic, optical, electromechanical, and electronic systems of satellite submillimeter radiometers have been worked out. Radiometers of the Obzor type are also intended for measurements of the thermal self-radiation of the upper layers of the earth's atmosphere [7].

The general appearance of the first modification of the instrument is shown in Fig. 6. The field of view of the radiometer's optical system is about 5°. A revolving mirror provides for the reception of radiation from the direction of the nadir and smooth deflections in the vertical plane provide for the measurement of the angular distribution of the brightness temperature of the radiation of the upper atmosphere, as well as for periodic calibrations against a thermal standard. The reception band is determined by the sensitivity of the indium antimonide receiver and is located between 300 and 2000 μ. The on-board nonnitrogen cryostat with a capacity of 10 liters provides for four to five days of functioning in orbital flight. A detailed description of the Obzor radiometer is contained in [7].

The material presented in this section indicates the complexity of the problem of creating on-board submillimeter radiometers and telescopes. In addition to the general difficulties in the creation of a scientific on-board apparatus a fundamental difficulty arises in the case of submillimeter instruments — providing prolonged deep cooling of the sensitive receiving elements and sometimes of the entire input unit under flight conditions.

3. A Program of Observations and the Basic Parameters for an On-Board Submillimeter Telescope

In the development of a submillimeter telescope it is necessary to carefully weigh what parameters it must possess for the solution of the scientific tasks set. It follows from the above presentation that these parameters can vary considey depending on the tasks set and on the potentialities of the objective for which the telescope intended.

In the planning of an on-board instrument for observis in the submillimeter range of the spectrum (from $\sim 50\ \mu$ to 2 mm) it is necessary first oíto single out the subjects which are of scientific interest. Then, allowing for the expected fluxes from the chosen subjects and the actual sensitivity of the radiation receivers, one must obtain estimates of the required telescope parameters — mirror size, field of view, threshold sensitivity, etc. After obtaining such estimates one can, conforming to the possibilities for mounting the instrument on board, select a reasonable mirror diameter (since it primarily determines the overall size and weight of the instrument) and then introduce corrections into the observation program, excluding subjects inaccessible for observations with the chosen mirror diameter and introducing new subjects.

a) Choice of Subjects for Study

The problems of extra-atmospheric submillimeter astronomy have been grouped in Section 1 as follows: 1) cosmological studies (relict radiation); 2) the study of line radiation of matter; 3) the study of bodies of the solar system; 4) the study of discrete sources of radiation outside the solar system; 5) aeronomical studies.

Such a division is dependent on differences in the physical nature of the observed radiation, and therefore the methods of studying the radiation differ. It actually does not seem possible to cover all five problems (even partially) with one instrument (see Section 1).

Below our attention will be concentrated mainly on the study of discrete sources of radiation outside the solar system; in addition, aeronomical studies and studies of planets of the solar system partially enter into the circle of our problems. Therefore the question is the creation of an on-board telescope with a mirror of the greatest possible size (i.e., with the best possible threshold sensitivity and resolving power) with very coarse spectral resolution.

As mentioned already, of the discrete sources of radiation the sources with unusual spectra for which the spectral density of the radiation increases upon passage into the submillimeter region are of the greatest interest for submillimeter astronomy. Composite spectra of some of the most studied sources are shown in Figs. 7 and 8 (the data are taken mainly from [39-43] for galactic sources and from [43-54] for extragalactic sources). They are divided into several diverse groups according to physical types.

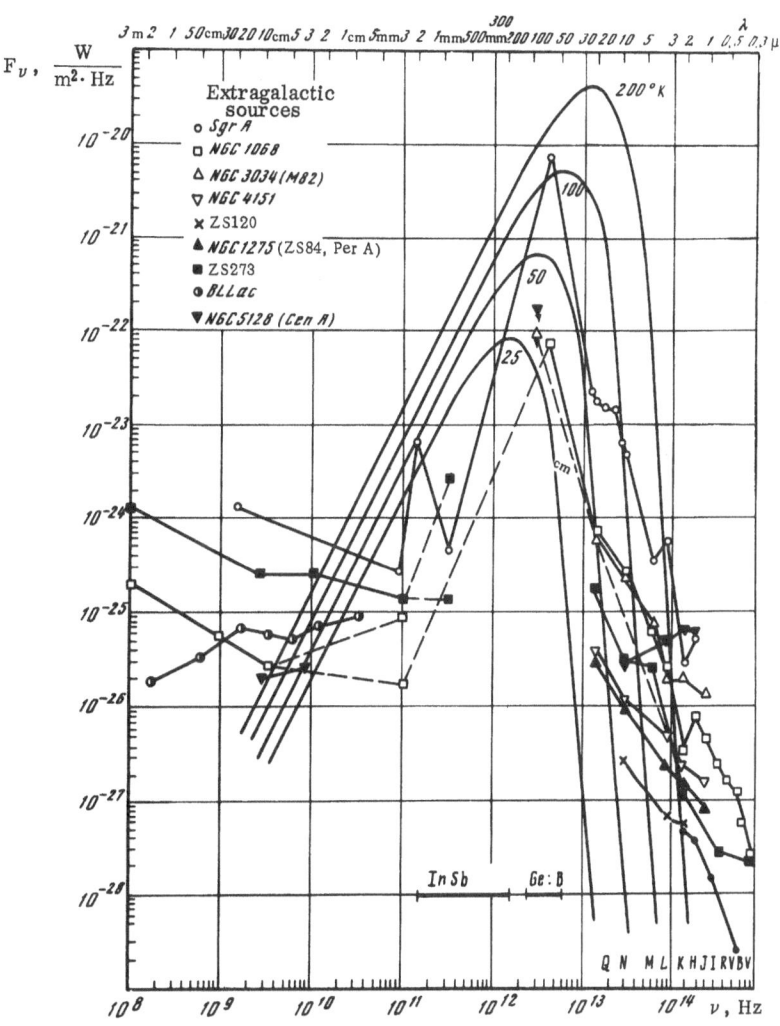

Fig. 7

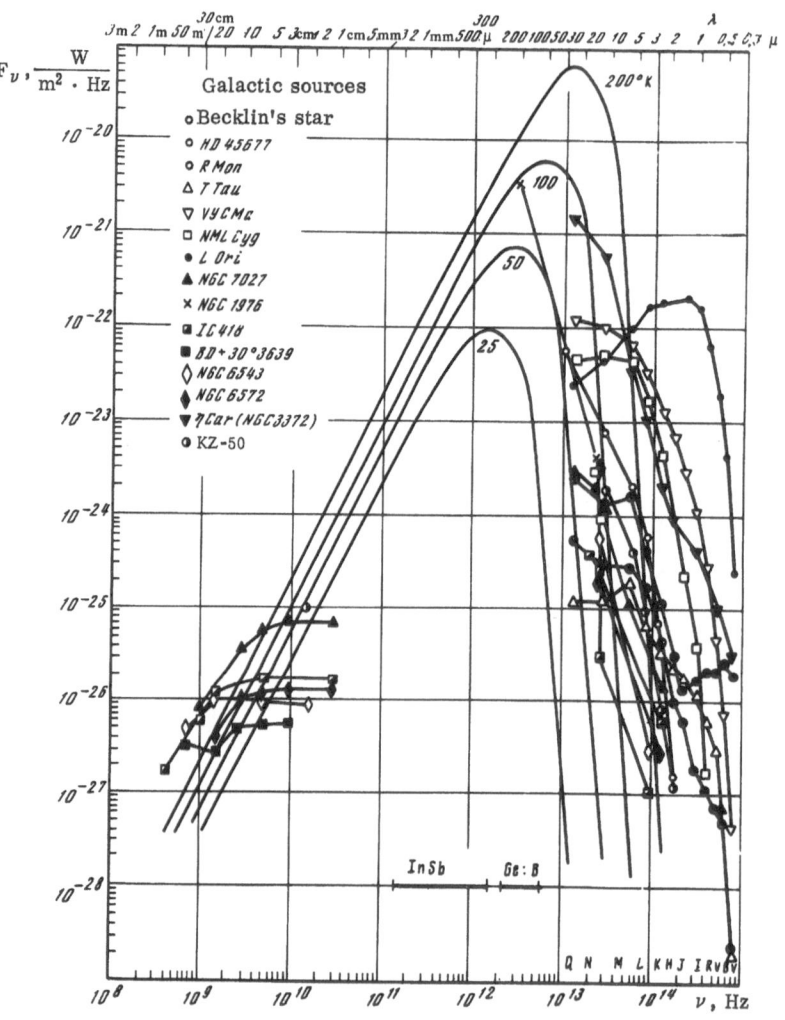

Fig. 8

Extragalactic Sources

1) galaxies (the center of our Galaxy Sgr A, Seyfert and N-galaxies NGC 1068, 1275, 3034, 4151, ZS120, the peculiar galaxy NGC 5128);

2) quasars (ZS273).

Galactic Sources

1) planetary nebulae (NGC 6543, 6572, 7027, BD+30°3639, IC418, KZ-50);

2) "infrared stars" (Becklin's star, HD45677, RMon, TTau, VYCMa, NMLCyg);

3) peculiar stars (η Car);

4) gaseous nebulae (NGC 1976).

Despite the diversity of types all these objects have similar infrared spectra with a characteristic steep rise in the spectral density of the radiation with an increase in wavelength from the visible region up to $\sim 20\ \mu$. The galactic and extragalactic sources differ somewhat in the nature of the spectra: For the galactic sources the rise in spectral density of the radiation begins (on the average) at shorter wavelengths and has a tendency toward saturation (for certain objects) in the region of $\sim 20\ \mu$. The available data for Sgr A and NGC 1068 in the region of $\lambda \approx 10\ \mu$ compel one to assume that for extragalactic sources one should expect a rise in the spectral density of the radiation upon passage into the submillimeter region. For the galactic sources observations in the submillimeter region should also give information on the spectrum.

The radiation spectra of absolutely black bodies with temperatures of from 25 to 200°K and with angular sizes of $\theta = 12''$ are plotted in Figs. 7 and 8 with thin lines (the angular size is chosen so as to match the scales). The close resemblance between the infrared spectra of all the sources and the spectra of black bodies at $T \approx 100\text{-}200°K$ leads to the thought that all these sources contain dense dust clouds having the corresponding temperatures. The absence of data in the submillimeter region prevents one from drawing an unambiguous conclusion concerning the nature of the infrared radiation of these objects. The fact that the spectrum of the "normal" cool giant Betelgeuse (α Ori) has a clearly expressed maximum at $\lambda = 1\ \mu$ and a decline in the longer-wave region, which is not so for "infrared stars," testifies in favor of the dust cloud hypothesis. For the planetary nebulae it has not yet been possible to connect the presence of an infrared excess with any of their other characteristics since "infrared" planetary nebulae differ but little from nebulae not having an infrared excess (except perhaps in the higher temperature of the central star).

Thus, the objects whose spectra are shown in Figs. 7 and 8 without doubt require study in the submillimeter region of the spectrum. Based on the trends in the variation of the spectra of these sources upon passage into the submillimeter region, we can conclude that for their reliable observation a telescope must have a threshold sensitivity of $10^{-24}\text{-}10^{-25}\ \text{W/m}^2 \cdot \text{Hz}$.

Studies of the planets of the solar system are most informative at a high spectral resolution (for the measurement of their atmospheres). With a coarse spectral resolution the observation of the planets presents less interest. Such planets as Mars and Jupiter have been observed in the submillimeter range [55, 56] and can serve as convenient calibration sources with fluxes (at $\lambda \approx 100\ \mu$) of $\sim 10^{-23}\ \text{W/m}^2 \cdot \text{Hz}$ (Mars) and $\sim 10^{-22}\ \text{W/m}^2 \cdot \text{Hz}$ (Jupiter).

In aeronomical studies (of the earth's atmosphere) the brightness temperature of the source has an order of 250°K [23]; the expected flux depends on the field of view and we will estimate it a little later.

b)   Threshold Sensitivity, Field of View,
and Size of Telescope Mirror

During observations of sources with small angular sizes θ_s, where $\theta_s < \theta$ (θ_s is the half-angular size of the source and θ is the half-angle of the telescope field of view), a telescope with a mirror of diameter D and area $A = \pi D^2/4$ receives from a source at the frequency ν a specific power

$$P_\nu = A\Omega_s B_\nu, \text{ [W/Hz]}, \tag{1}$$

where $\Omega = \pi \theta_s^2$ is the solid angle of the source and B_ν is the brightness of the source at the frequency ν (in $W/m^2 \cdot Hz \cdot sr$). When $\theta_s < \theta$ the value of $\Omega_s B_\nu = F_\nu$ does not depend on the instrument and represents the spectral flux density of the radiation from the source (in $W/m^2 \cdot Hz$). The optical system of the telescope provides to the receiver a specific power P_ν' equal to

$$P_\nu' = \eta P_\nu, \tag{2}$$

where η is the efficiency of the optical system, allowing for all possible losses.

From the entire spectrum of P_ν' the receiver records a signal in the band of its sensitivity from ν_1 to ν_2, where the effective power producing the electrical signal at the output of the receiver is equal to (in watts)

$$P' = \eta A \int_{\nu_1}^{\nu_2} F_\nu s(\nu)\, d\nu, \tag{3}$$

where $s(\nu) = S(\nu)/S_m$ is the normalized spectral sensitivity of the receiver; $S(\nu)$ is the volt-watt sensitivity of the receiver at the frequency ν; $S_m = S(\nu_m)$ is the maximum value of $S(\nu)$ at some frequency ν_m of the interval (ν_1, ν_2). For simplicity and clearness of the estimates we will henceforth assume that the source has a white spectrum in the interval (ν_1, ν_2), i.e., $F_\nu = F = const$ in (ν_1, ν_2). Then from (3) we find

$$P' = \eta A F \int_{\nu_1}^{\nu_2} s(\nu)\, d\nu = \eta A F \Delta\nu_{ef}, \tag{4}$$

where

$$\Delta\nu_{ef} = \int_{\nu_1}^{\nu_2} s(\nu)\, d\nu \tag{5}$$

is the effective sensitivity band of the receiver.

Now, by equating P' with the threshold power P_N of the receiver (which is determined with a measurement time constant $\tau = 1$ sec) we can find the threshold sensitivity of the telescope in units of the spectral flux density:

$$F_N = \frac{P_N}{\eta A \Delta\nu_{ef}}. \tag{6}$$

Using this equation one can construct the dependence of the threshold flux F_N on the mirror diameter D at a fixed efficiency η of the optical system and a certain receiver (P_N and $\Delta\nu_{eff}$). For definiteness we assumed that $\eta = 0.5$ and $P_N = 10^{-12}$ W, $\Delta\nu_{eff} = 10^{12}$ Hz. These figures are average for n-InSb and Ge:B receivers which are the most suitable for a submillimeter telescope (see Section 1 and [2]).

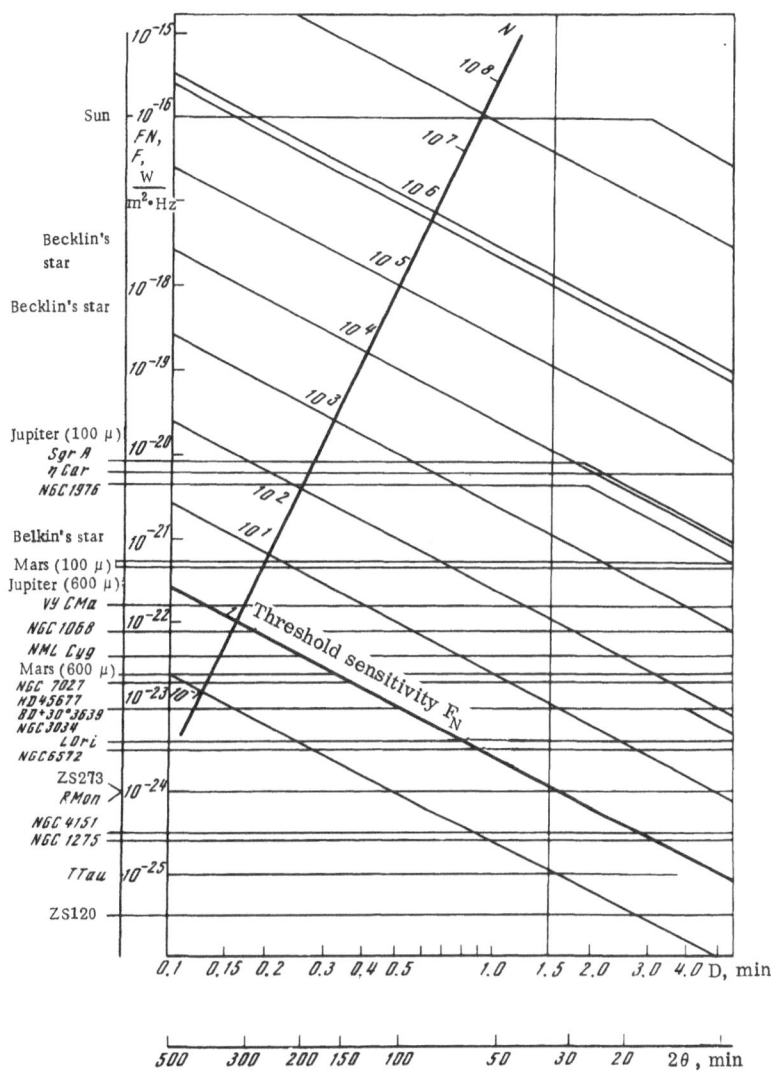

Fig. 9

With these values of the parameters the dependence of the threshold flux F_N (for a source of small angular size) on the mirror diameter D is represented in Fig. 9 by the straight line marked "threshold sensitivity." Using this dependence one can select the required mirror diameter for a given threshold flux or, conversely, find the minimum detectable flux for a given mirror diameter. As should be expected, the threshold sensitivity improves with an increase in mirror size.

We now introduce the signal-to-noise ratio N and, remembering that (6) is obtained for N = 1, we equate the effective power P' from (4) with the value NP_N. Then one can express the flux F from the source through N and the parameters of the telescope (η and A) and the receiver (P_N and $\Delta\nu_{eff}$)

$$F = \frac{NP_N}{\eta A \Delta\nu_{ef}}. \tag{7}$$

Lines of the dependence of F on D for different values of N are also plotted in Fig. 9. They are parallel to the line of threshold sensitivity N = 1; the scale of N is plotted on the inclined axis. From these data one can judge what signal-to-noise ratio we will obtain when observing a small source providing a flux F using a telescope with a diameter D. As is seen, for a given flux F the ratio N increases with an increase in the diameter D. Since the instrument's field of view 2θ decreases simultaneously in this case the increase in N will continue only so long as θ exceeds θ_s. As soon as θ is comparable with θ_s the increase in N ceases with a further increase in the diameter D. In order to clarify this let us now consider the case of the observation of an extended source.

When $\theta \leq \theta_s$ the specific power gathered by a mirror with an area A is

$$P_\nu = A\Omega B_\nu, \tag{8}$$

where, in contrast to (1), Ω_s is replaced by $\Omega = \pi\theta^2$. The magnitude of θ depends on D and consequently Ω also depends on D. Actually, by definition $\theta = y'/F$, where y' is the radius of the entrance window of the receiver and F is the focal distance of the telescope. If we introduce the value $f = F/D$ (the inverse relative aperture of the telescope) then we obtain $\theta = y'/fD$ and $\Omega = \pi(y'/fD)^2$. Since $A = \pi D^2/4$,

$$A\Omega = \frac{\pi D^2}{4}\frac{\pi y'^2}{f^2 D^2} = \left(\frac{\pi y'}{2f}\right)^2, \tag{9}$$

i.e., the product $A\Omega$ depends only on the size 2y' of the receiver entrance window and the inverse relative aperture f of the telescope, but not on the specific dimensions of the telescope. In the construction of Fig. 9 the value of f was fixed [$f = 2$ (F = 2D)] so that with variation in D the value of $A\Omega$ remained constant and equal to $1.38 \cdot 10^{-4}$ m$^2 \cdot$ sr for the chosen value y' = 15 mm.

Thus, for an extended source the power

$$P' = \eta\left(\frac{\pi y'}{2f}\right)\int\limits_{\nu_1}^{\nu_2} B_\nu s(\nu) d\nu \tag{10}$$

entering the receiver does not depend on the mirror size. If we again assume that the source has a white spectrum in the interval (ν_1, ν_2), i.e., $B_\nu = B = $ const, then we can express the signal-to-noise ratio for the given brightness B as

$$N = \frac{\eta B \Delta\nu_{ef}}{P_N}\left(\frac{\pi y'}{2f}\right)^2, \tag{11}$$

which also proves to be independent of the diameter D.

The fluxes of the sources of Figs. 7 and 8 extrapolated to $\lambda = 100 \mu$ are plotted in Fig. 9 on the left along the axis of fluxes F. The horizontal lines opposite these fluxes express on the inclined N scale the increase in N with an increase in D. As is seen, for some of the sufficiently large sources (the sun, Sgr A, NGC 1976) the increase ceases at the appropriate value of the telescope field of view.

Using Eq. (11) we can now find the value of N for the study of the earth's atmosphere, which in the submillimeter range has a brightness temperature $T_B \simeq 250°K$ [23]. With the receiver parameters chosen we obtain $N = 1.3 \cdot 10^6$ for $\lambda = 600 \mu$ and $N = 3.5 \cdot 10^7$ for $\lambda = 100 \mu$.

As is seen from Fig. 9, a mirror with a diameter of not less than 1 m is required for the observation of the majority of the interesting sources with the existing receivers. Based on the possibilities for the telescope arrangement it is advisable to select a maximum mirror diameter $D = 1.5$ m. Such a mirror diameter permits confidence in the reliable observation of a majority of the interesting discrete sources (if our extrapolation of the spectra to $\lambda = 100 \mu$ is correct). Calibration on board is possible from observations of Mars or Jupiter. As for the earth's atmosphere, for it the signal-to-noise ratio is sufficiently high to study the fine structure of the upper layers of the atmosphere.

c) Equipment of the Telescope

Having chosen a maximum mirror diameter $D = 1.5$ m we find that with a focal distance $F = 2D$ and a diameter $2y' = 30$ mm for the entrance window of the receiver the telescope will have a total field of view of $2\theta = 34'$. In order to carry out the program of Section 3 with such a field of view it is necessary to have the capability of aiming the telescope axis at a given point of space and keeping it in this direction with an accuracy on the order of 0.1θ, i.e., of ~ 2-$3'$. When it is impossible to accomplish such aiming the telescope should be equipped with an automatic scanning system.

In order to realize the threshold sensitivity of the receivers, as was noted above, the telescope must be equipped with cooled receivers (see [2]) and with low-noise amplifiers. For an estimate of the requirements on the input circuits of the preamplifiers let us turn to Table 2, where typical parameters are given for n-InSb and Ge:B receivers with which it is planned to operate the telescope.

The ranges of sensitivity, the threshold powers P_N, the volt–watt sensitivities S_m (at the maximum), the output voltages $\sqrt{e_N^2} = S_m P_N$ at the threshold signal, and the output resistances R_{out} are given in Table 2.

We will assume that the amplifier input coincides with the receiver output in noise ($R_{in} = R_{out}$). If one sets the condition that the noise voltage of the amplifier input must be no

TABLE 2

Characteristics of receiver	Ge : B	n-InSb
Range of sensitivity, μ	50—140	250—10000
Threshold power P_N, $W/Hz^{1/2}$	$4 \cdot 10^{-12}$	$4 \cdot 10^{-12}$
Sensitivity at maximum $S_{\nu m}$, V/W	$1.12 \cdot 10^8$	$5.8 \cdot 10^2$
Output voltage at threshold power $(\overline{e_N^2})^{1/2}$, V	$4.5 \cdot 10^{-8}$	$2.3 \cdot 10^{-9}$
Output resistance R_{out}, $k\Omega$	100	1
Noise temperature of preamplifier T_N at $R_{in} = R_{out}$, °K	360	95

greater than the noise voltage $\sqrt{\overline{e_N^2}}$ of the receiver then, using the Nyquist equation

$$\overline{e_N^2} = 4kTR_{in}\Delta f, \tag{12}$$

we obtain the values of the noise temperature

$$T_N = \frac{\overline{e_N^2}}{4kR_{in}\Delta f} \tag{13}$$

of the preamplifier input for each of the receivers. The corresponding values of T_N for Δf = 1 Hz (τ = 1 sec) are presented in the last row of Table 2 and show that the preamplifiers must be of very high quality (especially for n-InSb receivers).

The maximum overall coefficient of the amplifier is chosen from the condition that the noise voltage must be $\sim 1\%$ of the maximum output voltage at the amplifier output. Since the latter is chosen as 6 V, we have $G_{max} \simeq 3 \cdot 10^7$ for n-InSb and $1.5 \cdot 10^6$ for Ge:B.

It is seen from Fig. 9 that the dynamic range of the amplifier must be $\sim 10^4$ (from 10^{-24} to 10^{-20} W/m$^2 \cdot$ Hz) for the observation of discrete sources, which requires the use of logarithmic amplifiers or scaling division of the output voltages. For the study of the earth's atmosphere it is necessary to decrease the amplification another $\sim 10^4$ times in order to fit into the dynamic range of $\sim 10^4$. This can be accomplished with a stepped electrical or optical attenuator.

NOTE ADDED IN PROOF

In the time since the manuscript of the article was submitted a number of surveys and original articles have been published in which new apparatus is described and the results of further observations of the submillimeter radiation of discrete sources and the relict background are presented. The most important of these publications are presented below.

A. Collections and Surveys

1. Infrared Detection Techniques for Space Research. Proceedings of 5th ESLAB/ESRIN Symposium, D. Riedel Publ. (1972).
2. P. Thaddeus, "The short-wavelength spectrum of the microwave background," in: Ann. Rev. Astron. and Astrophys., 10 (1972).
3. M. Simon, Nature, 246, 193 (1973).

B. Original Reports

4. D. Y. Gezari, R. R. Joyce, and M. Simon, Astrophys. J., 179, 167 (1973).
5. I. Furniss, R. E. Jennings, and A. F. M. Moorwood, Nature, 236, 6 (1972).
6. P. Connes, J. Roy. Astron. Soc., 14, 288 (1973).
7. R. R. Joyce, D. Y. Gezari, and M. Simon, Astrophys. J., 171, L67 (1972).
8. B. T. Soifer, J. L. Pipher, and J. R. Houck, Astrophys. J., 177, 315 (1972).
9. H. M. Johnson, Astrophys. J., 180, L7 (1973).
10. J. P. Emerson, R. E. Jennings, and A. F. M. Moorwood, Nature, 241, 108 (1973).
11. E. F. Erickson, C. D. Swift, F. C. Witteborn, A. J. Mord, G. C. Angason, L. J. Carott, L. W. Kunz, and L. P. Giver, Astrophys. J., 183, 535 (1973).
12. D. A. Harper and F. J. Low, Astrophys. J., 182, L89 (1973).
13. G. H. Ricke, D. A. Harper, F. J. Low, and K. R. Armstrong, Astrophys. J., 183, L67 (1973).
14. J. P. Emerson, R. E. Jennings, and A. F. M. Moorwood, Astrophys. J., 184, 401 (1973).

LITERATURE CITED

1. A. S. Khaikin, Tr. FIAN, 77, 56 (1974).
2. A. A. Kobzev, V. I. Lapshin, S. V. Solomonov, and A. S. Khaikin, Tr. FIAN, 77, 80 (1974).
3. A. B. Fradkov and V. F. Troitskii, Tr. FIAN, 77, 85 (1974).
4. S. V. Solomonov, O. M. Stroganova, and A. S. Khaikin, Tr. FIAN, 77, 94 (1974).
5. V. N. Bakun, P. D. Kalachev, A. E. Salomonovich, and A. S. Khaikin, Tr. FIAN, 77, 103 (1974).
6. A. A. Kobzev, V. I. Lapshin, V. F. Troitskii, and A. S. Khaikin, Tr. FIAN, 77, 110 (1974).
7. A. E. Salomonovich, S. V. Solomonov, A. S. Khaikin, V. N. Gusev, and A. A. Kobzev, Preprint FIAN, No. 126 (1974).
8. A. E. Salomonovich, Usp. Fiz. Nauk, 99, 417 (1969); A. G. Kislyakov, Usp. Fiz. Nauk, 101, 607 (1970).
9. K. Shivanandan, J. R. Houck, and M. Harvit, Phys. Rev. Lett., 21, 1460 (1968); J. R. Houck and M. Harvit, Astrophys. J. Lett., 157, 145 (1969); J. L. Pipher, J. R. Houck, M. Harvit, and B. V. Jones, Nature, 231, 375 (1971); J. R. Houck, B. T. Soifer, M. Harvit, and J. L. Pipher, Astrophys. J., 178, L29 (1972).
10. A. G. Blair, F. Edeskuty, R. D. Hiebert, M. Jones, J. P. Shipley, and K. D. Williamson, Appl. Opt., 10, 1043 (1971); Phys. Rev. Lett., 27, 1154 (1971).
11. D. Muehlner and R. Weiss, Phys. Rev. Lett., 24, 742 (1970).
12. K. J. Johnston, S. H. Knowles, and P. R. Schwartz, Sky and Telescope, 44, 2 (1972).
13. R. L. Sorochenko, V. A. Pusanov, A. E. Salomonovich, and V. B. Shteinshleiger, Astrophys. J. Lett., 3, 7 (1963).
14. C. W. Tolbert, Nature, 206, 1304 (1965).
15. V. A. Efanov, A. G. Kislyakov, V. N. Kostenko, L. I. Matveenko, I. G. Moiseev, and A. P. Naumov, Izv. Vuz. Radiofiz., 12, 803 (1969).
16. J. E. Beckman, J. A. Bastin, and P. E. Clegg, Nature, 221, 994 (1969).
17. J. P. Oliver, E. E. Epstein, R. A. Shorn, and S. Lisoter, Astron. J., 72, 314 (1967).
18. L. I. Matveenko, Astron. Zh., 48, 1154 (1971).
19. P. Stettler, F. K. Kneubühl, and E. A. Müller, Astron. Astrophys., 20, 309 (1972).
20. I. L. Beigman, L. A. Vainshtein, and R. A. Syunyaev, Usp. Fiz. Nauk, 35, 267 (1968).
21. V. I. Lapshin, A. E. Salomonovich, S. V. Solomonov, V. F. Troitskii, A. B. Fradkov, and A. S. Khaikin, Izv. Vuz. Radiofiz., 13, 388 (1970).
22. M. Bater, R. M. Cameron, W. J. Burroughs, and H. A. Gebbie, Nature, 214, 377 (1967).
23. J. G. Beery, T. Z. Martin, J. G. Nolt, and C. W. Wood, Nature, 230, 36 (1971).
24. A. E. Basharinov, A. S. Gurvich, and S. T. Egorov, Dokl. Akad. Nauk SSSR, 188, 1273 (1969).

25. S. A. Zhevakin and A. P. Naumov, Izv. Vuz. Radiofiz., 6, 674 (1963).
26. P. Turon-Lacarrieu and J. P. Verdet, Ann. Astrophys., 31, 237 (1968).
27. J.-F. Moser, H. Steffen, and F. K. Kneubühl, Usp. Fiz. Nauk, 99, 469 (1969).
28. Techniques of Spectroscopy in the Far Infrared, Submillimeter, and Millimeter Regions of the Spectrum [Russian translation], D. Martin ed., Mir, Moscow (1970).
29. B. T. Ulrich, in: Low Temperature Physics, Proceedings of 12th International Conference, Academic Press (1971), p. 867.
30. V. S. Ivleva, A. A. Kobzev, V. I. Lapshin, A. E. Salomonovich, V. I. Selyanina, V. F. Troitskii, A. B. Fradkov, and A. S. Khaikin, Preprint FIAN, No. 12 (1971).
31. F. J. Low, H. L. Johnson, D. E. Kleinman, A. S. Latham, and S. L. Geisel, Astrophys. J., 160, 531 (1970).
32. H. H. Aumann and F. J. Low, Astrophys. J., 159, L59 (1970).
33. 36-Inch Airborne Infrared Telescope, NASA Ames Res. Center Rep. 9 (1971).

34.   N. J. Woolf, W. F. Hoffmann, C. L. Frederick, and F. Low, Science, 157, 187 (1967).
35.   W. F. Hoffmann and C. L. Frederick, Astrophys. J., 157, L9 (1969).
36.   W. F. Hoffmann, C. L. Frederick, and R. J. Emery, Astrophys. J., 164, L23 (1971).
37.   J. Gay, J. Lequex, J. P. Verdet, P. Turon-Lacrrieu, M. Bardet, J. Roucher, and Y. Zeau, Astrophys. Lett., 2, 169 (1968).
38.   Y. Terzian, Astrophys. Lett., 3, 87 (1969).
39.   N. J. Woolf, Astrophys. J., 157, L37 (1969).
40.   G. S. Khromov and V. I. Moroz, Astron. Zh., 48, 1122 (1971).
41.   R. H. Rubin and B. E. Turner, Astrophys. J., 157, L41 (1969).
42.   G. Neugebauer and G. Garmire, Astrophys. J., 161, L91 (1970).
43.   G. R. Burbidge and W. A. Stein, Astrophys. J., 160, 573 (1970).
44.   H. H. Aumann and F. J. Low, Astrophys. J., 161, L91 (1970).
45.   A. G. Kislyakov and A. P. Naumov, Astron. Zh., 44, 1324 (1967).
46.   M. M. Dworetsky, E. E. Epstein, W. G. Fogarty, and J. W. Montgomery, Astrophys. J., 158, L183 (1969).
47.   J. P. Hollinger, Astrophys. J., 142, 609 (1965).
48.   F. J. Low, D. E. Kleinmann, F. F. Forbes, and H. H. Aumann, Astrophys. J., 157, L97 (1969).
49.   E. E. Becklin and G. Neugebauer, Astrophys. J., 157, L31 (1969).
50.   F. J. Low, Astrophys. J., 159, L173 (1970).
51.   D. E. Kleinmann and F. J. Low, Astrophys. J., 159, L173 (1970).
52.   C. M. Wade, R. M. Hjellming, K. F. Kellermann, and J. F. C. Wardle, Astrophys. J., 170, L11 (1971).
53.   E. E. Becklin, J. A. Frogel, D. E. Kleinmann, G. Neugebauer, E. P. Ney, and D. W. Strecker, Astrophys. J., 170, L15 (1971).
54.   W. F. Hoffmann, C. L. Frederick, and R. J. Emery, Astrophys. J., 170, L89 (1971).
55.   H. H. Aumann, S. M. Gillespie, and F. J. Low, Astrophys. J., 157, L69 (1969).
56.   F. J. Low and S. W. Davidson, Astrophys. J., 142, 1278 (1965).

OPTICAL SYSTEMS OF ON-BOARD
SUBMILLIMETER TELESCOPES

A. S. Khaikin

Among the problems of submillimeter astronomy examined in [1], observations of discrete sources of radiation are of special interest. Such observations are possible only with the help of telescopes which focus the radiation of the source onto the entrance window of the receiver. Because of the absorption of submillimeter radiation in the earth's atmosphere the instruments must be placed on board extra-atmospheric flight apparatus. Here we will first of all examine some properties of the optical systems of on-board submillimeter telescopes; then, having selected a specific system answering our requirements, we will examine possible variants of its realization and give the method and results of the calculation of these variants.

1. SPECIFICS OF OPTICAL SYSTEMS OF ON-BOARD

SUBMILLIMETER TELESCOPES

The specifics of the optical systems of on-board submillimeter telescopes are determined by two circumstances which, while not having a fundamental nature, do lead to a situation which is unusual for optics and radiophysics.

First, practically the only type of sufficiently sensitive receivers for this range are photoresistors and bolometers cooled to the temperature of liquid helium (see [1], for example). The geometrical sizes of these receivers are considerably greater than the maximum wavelength of the submillimeter range ($\lambda \approx 1$ mm). Moreover, these receivers must be placed in helium cryostats, so that the radiation is brought to the receiver through a light pipe whose diameter (at a length of 200-300 mm required to produce the minimum admissible heat influx) cannot be made less than ~10 mm. The diameter of the entrance window of the light pipe usually reaches 20-30 mm. Here the angular aperture of the beam of radiation incident on the entrance window must be small enough (10-20°) to reduce the number of reflections in the light pipe and thus losses. Therefore the specifics of the receivers, by not permitting one to obtain images whose size would be limited only by diffraction, leads to the softening of the demands on the focusing of the received radiation, i.e., on the quality of the optical surfaces, and in addition compels one to construct very long-focus systems.

Second, the necessity of raising the telescopes into the top layers or even beyond the limits of the earth's atmosphere greatly complicates the solution of the problem of the exact aiming and keeping of the optical axis of the telescope on the source studied. In the majority of cases the optical axis of a submillimeter on-board telescope cannot yet be stabilized with an accuracy better than several angular minutes. This in turn places a limitation on the attainable field of view (the directional diagram) of the telescope and prevents it from being less than 10-20', as a consequence of which the possibilities for spatial resolution are reduced.

A. S. KHAIKIN

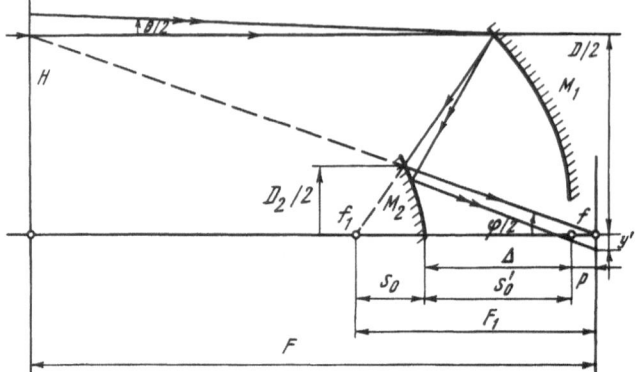

Fig. 1. Diagram of Cassegrain telescope objective.

When the methods of reception of radiation developed in radiophysics (superheterodynes, high-frequency low-noise amplifiers) are sufficiently mastered in the submillimeter range that it proves possible in practice to construct on-board instruments on their basis, the demands on the optics of the latter will prove to be just as rigid relatively as in the radio range. This will mean that the deviations from the theoretical surface of the mirrors must comprise small fractions of the minimum wavelength of the range while the size of the focal spot will be limited by diffraction effects at the edges of the mirrors. On the other hand, this requres a sharp increase in the aiming accuracy. For the present the forced limitations mentioned above to a certain extent permit the easing of the task of creating telescopes for the submillimeter range.

Let us examine in more detail the characteristic parameters of on-board submillimeter telescopes.

With the sizes indicated above for the entrance window (20-30 mm) and the field of view (~20') it is easy to see that the focal distance of the telescope should be 2500-5000 mm. Here we encounter on-board restrictions of another kind — restrictions in the dimensions of the instruments — which lead us to the problem of constructing as short an optical system as possible with a large enough focal distance.

The demands on the image quality are determined by the size of the entrance window of the receiver: The focal spot from an infinitely distant point must not be larger than the entrance window. A considerable decrease in the focal spot in comparison with the entrance window does not provide any advantages but only leads to the very complex problem of matching the receiver with the small entrance aperture of the light pipe and to an increase in reflection losses in the light pipe.* Thus, the optical system can possess considerable aberrations. We note that diffraction effects do not play the decisive role in this case since for the parameters indicated above and $\lambda = 1$ mm the diameter of the central diffraction spot is equal to the diameter of the entrance window for a very moderate diameter of the system on the order of 150-170 mm. Of course, for work in a wide range of the spectrum the system should be achromatic, and consequently constructed from reflecting elements.

All the requirements listed are satisfied by a single system — the reflecting Cassegrain telescope objective, a diagram of which is shown in Fig. 1. It consists of a concave main

---

* The questions of matching the receiver with the light pipe entrance are examined in [2].

mirror M_1 and a convex secondary mirror M_2 located in front of the focus f_1 of mirror M_1 and displacing the image of an infinitely distant point from f_1 to f where the focus of the system lies. The principal plane H of the system is carried far forward, thanks to which the focal distance F of the system is considerably greater than the focal distance F_1 of the main mirror M_1 or the length Δ of the system. The Cassegrain telescope objective represents a prefocal system (the secondary mirror is located in front of the focus of the main mirror) and therefore always has a smaller length Δ (for the same focal distance F) than a Gregorian postfocal system with a concave secondary mirror.

With a spherical shape for both mirrors a Cassegrain objective always has an undercorrected spherical aberration: The positive spherical aberration of the main mirror, magnified in the secondary mirror, predominates over the negative spherical aberration of the secondary mirror. In this case the dispersion spot in the Gaussian plane is rather large: In front of the Gaussian plane, however, there is a spot of least dispersion with a relatively small size. In Section 3 below we will show that for main mirror apertures which are not very large one can bring the spot of least dispersion into a given plane by the choice of the focal distance of the secondary mirror, with the size of the spot being fully satisfactory from the point of view of our problem.

Stigmatism at the axis in a Cassegrain telescope objective can be achieved in two ways (see [3], for example): either by compensation through an aspherical shape for one or both mirrors or by the use of a parabolic main mirror and a hyperbolic secondary mirror, each corrected separately. The fabrication of aspherical mirrors is necessary in both cases; in a compensated telescope objective the main mirror can be made spherical, but for large sizes (a diameter of 1000-1500 mm) it is complicated to make. The difficulties are aggravated by the fact that the main mirror of a large telescope must have a large relative aperture to obtain the smallest possible length of the system. In Section 3 we will examine a variant of a large telescope with spherical mirrors and in Section 4 a variant using a ready-made searchlight mirror 1500 mm in diameter of approximately parabolic form.

We note that another approach to the optical systems of submillimeter telescopes is possible. The point is that in order to increase the signal-to-noise ratio the aperture of the beam irradiating the cooled photoresistor must be as large as possible. In placing the receiver in a cryostat the problem of transforming the narrow beam from the optical system into a wide beam incident on the receiver is solved with the help of an immersion lens or an integrating chamber mounted at the "cold" end of a long light pipe (the light pipe is made long to decrease the inflow of heat to the receiver from outside). The losses in the light pipe are slight because of the small aperture of the beam coming from the long-focus optical system. One can, however, consider the photoresistor by itself without relation to the method of cooling. Then it turns out that the optical system of the telescope should have the maximum exit aperture. A calculation for such a system with a diameter of 1000 mm having a full exit aperture of 60° with a receiver radius of 1 mm is presented in [4]. Both mirrors are aspherical in shape: The main mirror is a very deep paraboloid changing asymptotically into a cone while the secondary mirror is a slightly deformed cone with an angle of 45° at the apex. The fabrication of such a system is very problematical; in addition, the problem of cooling and mounting the receiver is completely ignored in this case.

2. PARAXIAL CALCULATION OF CASSEGRAIN REFLECTING

TELESCOPE OBJECTIVE

In order to facilitate the subsequent presentation let us give the basic equations and the method of the paraxial calculation of a Cassegrain telescope objective and explore some of its possibilities. In the nomenclature and terminology we will follow Maksutov [3].

The telescope objective (Fig. 1) is characterized by the position parameters α and β:

$$\alpha = F_1/s_0, \tag{1}$$

$$\beta = s_0/s_0', \tag{2}$$

where F_1 is the focal distance of the main mirror and s_0 and s_0' are the front and back segments of the secondary mirror. It can be found from Fig. 1 that $F_1 = \beta F$ and $D/D_2 = \alpha$, where F is the focal distance of the system. Thus, α characterizes the approximate ratio of the mirror diameters while β characterizes the elongation of the focal distance of the telescope objective compared with the focal distance of the main mirror.

For the calculation of the telescope objective one must first of all assign the diameter D of the main mirror and the working segment p, i.e., the distance from the apex of the main mirror to the Gaussian plane f of the telescope objective, where the entrance window of the receiver of the assigned diameter 2y' is located. Then one must assign any two of the three parameters F, F_1, and Δ (the length of the system), depending on what considerations are started from in the calculation. If the focal distances of the main mirror and of the entire system F_1 and F are fixed then we find

$$\beta = F_1/F, \tag{3}$$

$$s_0' = (F_1 + p)/(\beta + 1), \tag{4}$$

$$\Delta = s_0' - p. \tag{5}$$

If the main mirror and the length of the system F_1 and Δ are fixed then we have

$$s_0' = \Delta + p, \tag{3'}$$

$$\beta = (F_1 - \Delta)/s_0', \tag{4'}$$

$$F = F_1/\beta. \tag{5'}$$

Finally, if the focus and the length of the system F and Δ are fixed then the following equations are obtained:

$$s_0' = \Delta + p, \tag{3''}$$

$$\beta = \Delta/(F - s_0'), \tag{4''}$$

$$F_1 = \beta F. \tag{5''}$$

The subsequent calculation in all three cases is carried out identically: We find the front segment of the secondary mirror

$$s_0 = \beta s_0', \tag{6}$$

the ratio of mirror diameters

$$\alpha = F_1/s_0, \tag{7}$$

the diameter of the secondary mirror

$$D_2 = D/\alpha \tag{8}$$

and its focal distance

$$F_2 = -s_0/(1 - \beta). \tag{9}$$

Then we calculate the geometrical field of view of the system

$$2\theta = 2y'/F \tag{10}$$

and the coefficient of screening of the area of the main mirror by the secondary mirror

$$\eta_A = (D_2/D)^2 = 1/\alpha^2. \tag{11}$$

The size calculation of the telescope objective comes down to the calculation of the increment ΔD_2 in the diameter of the secondary mirror due to the finite field of view of the system:

$$\Delta D_2 \cong 2\theta\Delta \sqrt{1 + A_1^2/4}. \tag{12}$$

A somewhat different method can be suggested for rapid estimates of the parameters and possibilities of the telescope objective during its construction. We introduce in addition to the parameters α and β three more dimensionless parameters which are important in a constructional respect: the ratio of the length of the system Δ to its focal distance F:

$$\gamma = \Delta/F, \tag{13}$$

the ratio of the working segment p to the focal distance of the main mirror F_1:

$$\varkappa = p/F_1 \tag{14}$$

and the ratio of the focal distances of the secondary and main mirrors

$$\mu = -F_2/F_1. \tag{15}$$

Now, using the equations of the paraxial calculation, one can obtain for the five parameters three universal equations which are satisfied by all Cassegrain telescope objectives:

$$\gamma = \beta\frac{\alpha - 1}{\alpha}, \tag{16}$$

$$\varkappa = \frac{1 - \beta(\alpha - 1)}{\alpha\beta}, \tag{17}$$

$$\mu = \frac{1}{\alpha(1 - \beta)}. \tag{18}$$

Since the Cassegrain telescope objective belongs among the so-called elongating systems [3] the parameters β and γ for it are limited to the interval $(0, 1)$. The range of variation in the parameter α is dictated by the following considerations. The choice of $\alpha < 2$ leads to excessive screening of the area of the main mirror and to an increase in the role of diffraction at the round opening of the objective. When $\alpha > 5$ the relative aperture of the secondary mirror is extremely large, which complicates its fabrication. Hence for α one should be confined to the interval $2 \leq \alpha \leq 5$.

Graphs of Eqs. (16)-(18) for the indicated ranges of variation of the parameters are given in Fig. 2. The dependences of γ, $\varkappa$, and μ on β with $\mathring{\alpha}$ as the parameter are shown here. The dependence of γ on β (Fig. 2a) illustrates the effect of elongation of the focus: The telescope objective is always shorter than its focal distance ($\gamma < 1$), with the values of γ and β differing little (by a maximum of twofold when $\alpha = 2$). To obtain a relatively short working segment p in a system with a considerably elongated focus (small values of β) it is necessary to choose α as large as possible, as the dependence of $\varkappa$ on β shows (Fig. 2b). At the same time, as seen from the graphs of $\mu = f(\beta)$, an increase in α leads to an increase in the curvature of the secondary mirror (Fig. 2b). Therefore a compromise solution must be found in the choice of α.

We can demonstrate the use of these graphs in rapid estimates of the parameters of a system on a concrete example (it is shown by arrows in Fig. 2). We assume that it is necessary to estimate a system with a relative length $\gamma = 0.2$. We select the value $\alpha = 4$ which is

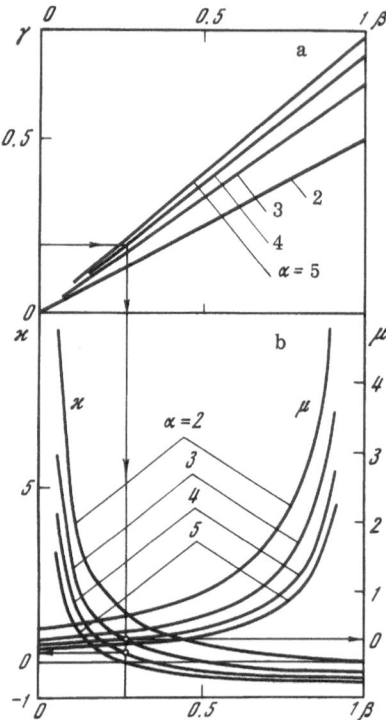

Fig. 2. Graphs of universal relationships between dimensionless parameters of a Cassegrain telescope objective.

not too large. We have (Fig. 2a) $\beta \simeq 0.26$. Extending the arrow to Fig. 2b, for $\alpha = 4$ we find
from its intersection with the $\varkappa$ and μ curves that $\varkappa \simeq 0.2$ and $\mu \simeq 0.35$. Now it is sufficient
to select from the constructional considerations (the placement of the entrance window of
the receiver) a value for the working segment p, such as 200 mm, and using $\varkappa$, β, and μ we
obtain the values $F_1 = 1000$ mm, $F = 3850$ mm, $F_2 = -350$ mm, and $\Delta = F_1(\alpha - 1)/\alpha = 750$ mm.
Having assigned the field of view $2\theta = 20' = 1/172$ rad, we obtain the required diameter of the
entrance window of the receiver, $2y' \simeq 22$ mm. The diameter of the main mirror in our case,
in contrast to radio astronomical and optical telescopes, is chosen not from considerations of
resolving power but from the area of the telescope needed to achieve a given level of receiver
sensitivity (see [1]). A mirror with a diameter of not less than 1000 mm is necessary for the
observation of discrete sources in the submillimeter range with the existing receivers [1].
Let us take D = 1000 mm in our concrete example ($A_1 = 1 : 1$); then we find $D_2 = 250$ mm ($A_2 =
1 : 1.4$) and $\Delta D_2 \simeq 5$ mm. Thus, we have estimated the parameters of a certain optical system;
corrections can now be introduced (if they are needed) and the method of calculation from
Eqs. (1)-(12) can be used for an exact calculation of the paraxial parameters.

The paraxial calculation described is suitable for telescope objective mirrors of any
shape. In the case of aspherical mirrors the focal distances F_1 and F_2 should be understood
to be the apex focal distances $F = \mathring{R}/2$, where $\mathring{R}$ is the radius of curvature of the respective
mirror at the apex. A paraxial calculation, however, does not allow one to obtain any information concerning the quality of the image.

The aberration calculation of a telescope objective presents a special problem and requires taking into account the conic section of the beams, and we will examine it in Section 3,

but here we present only the result which shows how the use of aspherical mirrors achieves the correction for spherical aberration [3]. For this the eccentricities e_1 and e_2 of the mirrors must satisfy the equation

$$e_2^2 = \frac{(1 - \beta)(1 + \beta)^2 + \alpha(1 - e_1^2)}{(1 - \beta)^3}. \tag{19}$$

For example, if the main mirror is made parabolic ($e_1^2 = 1$) then $e_2^2 = (1 + \beta)^2/(1 - \beta)^2 > 1$, i.e., the secondary mirror must be hyperbolic. This is the classical variant of the Cassegrain telescope objective in which each of the mirrors is corrected independently for spherical aberration.

For a spherical main mirror ($e_1^2 = 0$) it is necessary that

$$e_2^2 = \frac{(1 - \beta)(1 + \beta)^2 - \alpha}{(1 - \beta)^3}.$$

Since $\alpha < 2$ this means that $e_2^2 < -1$, i.e., the secondary mirror must be an oblate spheroid.

Finally, if the secondary mirror is made spherical then for the main mirror we find

$$e_1^2 = 1 - \frac{(1 - \beta)(1 + \beta)^2}{\alpha},$$

which denotes an elliptical shape. The latter two variants illustrate the compensation method of correcting for spherical aberration.

3. CASSEGRAIN TELESCOPE OBJECTIVE

WITH SPHERICAL MIRRORS

The fabrication of aspherical optics is connected with certain difficulties and is not always accessible. At the same time, as we have seen, in telescopes of the submillimeter range the rather considerable blurring of the image of an infinitely distant point source is admissible. Therefore there is undoubted interest in the study of the aberrational properties of a reflecting Cassegrain telescope objective constructed from components of spherical shape. This is especially important for telescopes of moderate diameter (up to ~200-300 mm) since such spherical mirrors of quite good quality can be fabricated in almost any optical shop. This is also interesting for larger telescopes since the fabrication and testing of a spherical mirror is always easier than for an aspherical mirror.

For the calculation of the spherical aberration of a telescope objective one can use a general expression for the longitudinal spherical aberration $\Delta s'$ of the mirror surface which is of second order with the eccentricity e [3]:

$$\Delta s' = y^2 \frac{s^2 e^2 - (s + 2F)^2}{8F(s + F)^2}, \tag{20}$$

where y is the radius of the mirror zone (the height of the point of intersection of the ray with the mirror relative to the optical axis), s is the front segment (the distance from the mirror to the point of intersection of the ray with the axis), and F is the focal distance of the mirror. In the case of a mirror of spherical shape one must set $e^2 = 0$ in (20).

The aberration Δs_1 of a spherical main mirror is obtained from (20) in the limiting transition to $s \to \infty$ (a source of infinite remoteness):

$$\Delta s_1 \simeq -y_1^2/8F_1, \tag{21}$$

where y_1 is the height of the ray above the axis. For a spherical secondary mirror the aberration Δs_2 is obtained from (20) by direct substitution of the corresponding parameters (since for it the source is located at a finite distance s_0):

$$\Delta s_2 = -y_2^2 \frac{(s_0 + 2F_2)^2}{8F_2(s_0 + F_2)^2} \simeq \frac{y_1^2}{8\beta^2 F_1} \frac{(1 - \beta)(1 + \beta)^2}{\alpha}. \tag{22}$$

An inaccuracy is allowed in the latter equation to simplify the estimates: y_2 is replaced by y_1/α. This is correct only in the absence of aberration of the main mirror. Later in the exact calculation of aberrations we will use the exact relationship between y_1 and y_2, but here we leave it approximate.

The longitudinal magnification of the secondary mirror is $q_2 \simeq -1/\beta^2$ (the equation is exact for small angles between the rays and the axis and approximate for large angles). Therefore the aberration of the main mirror magnified by the secondary mirror will equal

$$\Delta s_1' = q_2 \Delta s_1 \simeq y_1^2/8\beta^2 F_1. \tag{23}$$

Since $\Delta s_1'$ and $\Delta s_2'$ are opposite, the total limiting spherical aberration of the telescope objective is expressed by their difference

$$\Delta s' = \Delta s_1' - \Delta s_2 \simeq \frac{y_1^2}{8\beta^2 F_1}\left[1 - \frac{(1 - \beta)(1 + \beta)^2}{\alpha}\right]; \tag{24}$$

it represents (Fig. 3) the distance from the Gaussian plane to the point of intersection with the axis of the ray passing through the zone y_1 of the main mirror. We consider the value of $\Delta s'$ as positive if the point of intersection of the beam with the axis lies in front of the Gaussian plane.

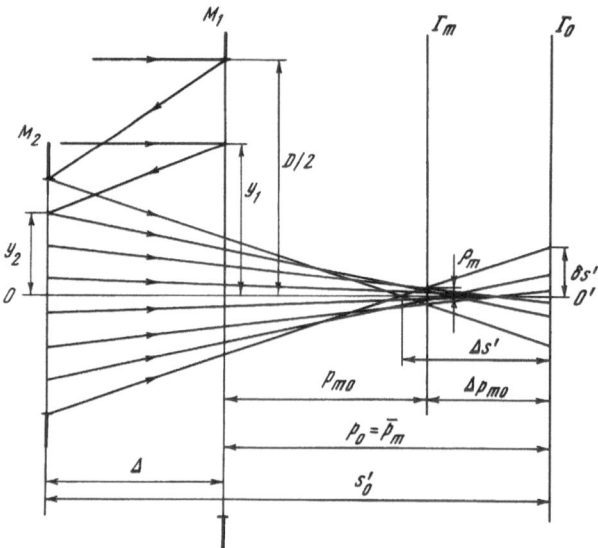

Fig. 3. Spherical aberration and the spot of least dispersion for a Cassegrain telescope objective with spherical mirrors. M_1 and M_2 are the apex planes of the main and secondary mirrors, respectively; OO' is the optical axis.

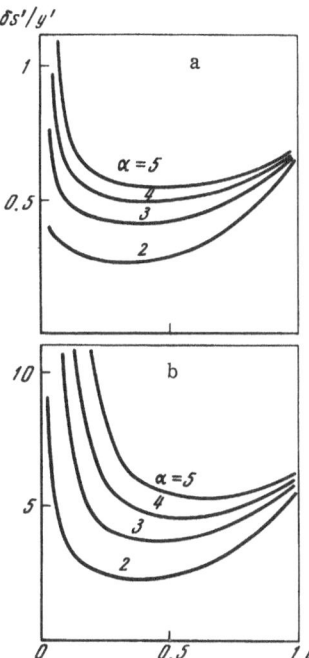

Fig. 4. Transverse spherical aberration in the
Gaussian plane of Cassegrain telescope objectives with spherical mirrors, normalized to the
radius of the entrance window of the receiver
for a total field of view $2\theta = 20'$. a) Relative
aperture of main mirror $A_1 = 1{:}2$; b) $A_1 = 1{:}1$.

If one considers the ray (Fig. 3) entering the system through the edge of the main mirror
($y_1 = D/2$) and seeks the value of the transverse spherical aberration of this ray in the Gaussian
plane Γ_0 then by simple transformations one can find for the total transverse aberration $\delta s'$
normalized to the radius of the entrance window of the receiver the following:

$$\frac{\delta s'}{y'} \simeq \frac{3A_1}{2\theta}\left\{\frac{A_1^2}{32\beta}\left[1 - \frac{(1-\beta)(1+\beta)^2}{\alpha}\right]\right\}\Big/\left\{1 - \alpha\frac{A_1^2}{32\beta}\left[1 - \frac{(1-\beta)(1+\beta)^2}{\alpha}\right]\right\}, \tag{25}$$

where $A_1 = D/F_1$ is the relative aperture of the main mirror. The dependence of $\delta s'/y'$ on β
for different values of α at $2\theta = 20'$ is shown in Fig. 4a ($A_1 = 1{:}2$) and Fig. 4b ($A_1 = 1{:}1$).

As seen from Fig. 4a, for the moderate value $A_1 = 1{:}2$ the spot of spherical aberration
in the Gaussian plane falls entirely within the size of the entrance window of the receiver for
all reasonable values of β (larger than ~0.1) and α. Therefore, by adopting mirrors with
$A_1 \le 1{:}2$, telescope objectives which are satisfactory with respect to the size of the focal spot
could be constructed entirely from spherical components without any measures for decreasing
the spherical aberration. However, the striving toward the smallest possible relative length
γ of the system compels us to increase A_1 to $1{:}1$ or more. As Fig. 4b shows, the increase in
A_1 to $1{:}1$ increases the aberration spot in the Gaussian plane by an order of magnitude so that
its size (for small β) exceeds the size of the entrance window of the receiver by 5-10 times.
This fact does not at all mean that it is impossible to construct a telescope objective of spherical components which has a field of view of $2\theta = 20'$ and $A_1 = 1{:}1$. The point is that because
of the approximations adopted the estimates of $\delta s'/y'$ from Eq. (25) prove to be somewhat over-
stated at large values of A_1. In addition, as is known, the structure of a beam distorted by
spherical aberration is such that outside the Gaussian plane Γ_0 (in front of this plane in the

given case, when the system is undercorrected and $\Delta s'$ is positive) there is a plane Γ_m with a spot of least dispersion whose diameter can be several times smaller than the value $\delta s'/y'$ (Fig. 3).

The simplest solution to the problem is to refocus the system, i.e., to move the receiver (or its entrance window) to the plane Γ_m. However, an analytical calculation of the distance p_{m_0} to this plane and of the radius ρ_m of the spot of least dispersion does not appear to be practicable for complicated systems. Therefore it is necessary to undertake numerical calculations in order to find the values of p_{m_0} and ρ_m. In addition, the moving of the receiver or its entrance window to in front of the main mirror is not always possible based on constructional considerations. In such a case it is necessary not only to calculate the coordinate p_{m_0} and the size ρ_m of the spot of least dispersion but also to correct the parameters of the telescope objective so that this spot is at a given distance $\bar{p}_m = p$ from the main mirror.

Thus, the problem is stated as follows. Having calculated the initial paraxial parameters of the telescope objective, one must find the coordinate p_{m_0} and the size ρ_m of the spot of least dispersion. In this case the Gaussian plane Γ_0 lies at a distance $p_0 = p$ from the main mirror. Then one must shift Γ_0 to the right by a change, for example, in the focal distance F_2 (and consequently in β and F) to some new position p_1 and again calculate the parameters p_{m_1} and ρ_{m_1} of the spot of least scattering. If $p_{m_1} \neq \bar{p}_m$ (with the assigned accuracy) then one must again shift Γ_0 (a new position p_2) and calculate p_{m_2} and ρ_{m_2}, and so forth until it is found that $p_{m_i} = \bar{p}_m$ with the assigned accuracy. Of course, with such a correction the size ρ_m of the spot increases somewhat and it must be determined that it has not gone beyond the admissible bounds, or else the parameters of the system must be changed so as to decrease ρ_m. As the step in the shifting of the Gaussian plane during the correction of the parameters it is natural to use the difference between the value of p_m obtained in the i-th stage and the assigned value (Fig. 3), i.e.,

$$\Delta p_{mi} = p_{mi} - \bar{p}_m, \qquad p_{i+1} = p_i - \Delta p_{mi}. \tag{26}$$

The conditions of solvability of such a problem can be roughly found as follows. It is necessary that during the shifting of the Gaussian plane by a distance Δ_p (by a change in β) the point of intersection of the edge ray with the axis is shifted by a distance Δ_m of the same order (in this lies the approximate nature of the estimates: we follow the variations in p_m for the edge ray). For the estimate one can assume that

$$\Delta_p \approx \frac{ds_0'}{d\beta} = -\frac{F_1}{\alpha\beta^2},$$

$$\Delta_m \approx \frac{d(\Delta s')}{ds} \simeq -\frac{2y_1^2}{8\beta^3 F_1}\left[1 - \frac{(1-\beta)(1+\beta)^2}{\alpha}\right] + \frac{y_1^2(1+\beta)}{4\alpha\beta F_1}.$$

For $y_1 = D/2$ (the edge ray) we find from the condition $\Delta_m \gtrless \Delta_p$ that

$$A_1 \leqslant 4\sqrt{\frac{\alpha - 1 + 2\beta^3 - \beta}{\beta}}.$$

Since $\beta \ll 1$ in long-focus systems one can neglect $2\beta^3 - \beta$ in comparison with $\alpha - 1$ under the radical and obtain

$$A_{1\,ex} \leqslant \frac{4\sqrt{\alpha\gamma}}{\alpha - 1}. \tag{27}$$

The dependence of $A_{1\,ex}$ on γ for different α is shown in Fig. 5. As is seen, the problem is solvable in principle for $A_1 \approx 1:1$ when $\gamma \gtrless 0.1$. Unfortunately, one cannot obtain reasonable estimates for the radius ρ_m and the solution of the problem of its value must await concrete calculations.

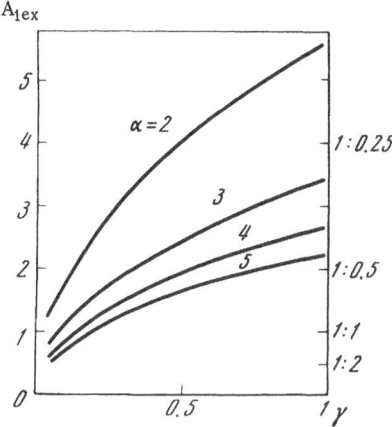

Fig. 5. Extreme relative apertures $A_{1\,ex}$ of the main mirror for which the problem of establishing the spot of least dispersion in the assigned plane is still solvable.

We have compiled a program of calculations on M-220 and Nairi-2 computers for the calculation and correction of the parameters of a telescope objective. The calculations are conducted according to the following system. First of all the initial paraxial parameters of the system are calculated from Eqs. (1)-(12). Then the main mirror is divided into 25 equal annular zones and the coordinates (angles and points of intersection with the axis) of 50 rays emerging from the system and corresponding to these zones in the meridional plane are calculated. Here the aberrations of the rays are calculated from equations analogous to (21) and (22) but without the inaccuracies allowed there.

The coordinate p_{mi} of the plane Γ_m of least dispersion and the radius ρ_{mi} of the spot in this plane are sought on the basis of the coordinates of the 50 rays. Then the new distance to the Gaussian plane is found in accordance with (26). Then the new parameters (β, F_2, F, etc.) are calculated from the value of p_{i+1} and Eqs. (3'), (4'), (5'), and (6)-(12), and the cycle is repeated until the condition $|p_{mi} - \bar{p}_m| < \varepsilon$ is satisfied, where ε is the assigned accuracy. Here the correction is considered as finished and the parameters of the system obtained are taken as the final parameters.

The effectiveness of the method described can be demonstrated on the example of the calculation of the telescope objective for an instrument intended for the study of the sun's spectrum in the submillimeter range [5]. The initial data for the calculation were the following: diameter of entrance window $2y' = 11$ mm, required field of view $2\theta \approx 40'$ (with the sun's angular diameter of 30'), so that the focal distance of the system should be $F \approx 800-1000$ mm. From constructional considerations we chose: working segment p = 290 mm, diameter and focal distance of main mirror (it is fixed) D = 160 mm, $F_1 = 140$ mm ($A_1 = 1:0.87$), length of system $\Delta = 70-80$ mm. Three variants $\Delta = 70$, 75, and 80 mm were calculated and corrected from Eqs. (3')-(5') for the fixed F_1 and Δ. The results are presented in Table 1 (here, in addition to the nomenclature already known, is given the value x, the distance from the plane of least dispersion to the Gaussian plane; all the dimensional parameters are in millimeters). As seen from Table 1, in all three variants the spot of least dispersion has a fully acceptable diameter $2\rho_m = 4-6$ mm, which is $\sim 2-3$ times less than the diameter of the entrance window. If the parameters of one of the variants, such as $\Delta = 70$ mm, are subsituted into Eq. (25) then we obtain a value of ~ 2 for $\delta s'/y'$ in the Gaussian plane. By placing the receiver in the Gaussian plane of the telescope objective we could not have obtained the desired field of view and would have lost instrument sensitivity. Of the three variants of Table 1 the variant $\Delta = 75$ mm

TABLE 1. Variants of Calculation and Correction of
Cassegrain Telescope Objective for Study of the Sun

Param-eter	$\Delta = 70$	$\Delta = 75$	$\Delta = 80$	Param-eter	$\Delta = 70$	$\Delta = 75$	$\Delta = 80$
F	846	976	1158	F_2	−84	−76	−68
β	0.165	0.144	0.121	D_2	80	74	68
α	2.00	2.15	2.33	x	63	88	126
$2\rho_m$	4.2	5.0	6.0	$2\theta,'$	45	39	33

Parameters common to the three variants: $F_1 = 140$, $D = 160$, $2y' = 11$, $\bar{\rho}_m = 290$.

was put into practice [5]. The telescope objective prepared fully confirmed the calculations: The diameter of the sun's image was equal to about 12 mm compared with an expected 13 mm (a geometrical image of ~8 mm and aberrational spreading of ~5 mm).

Let us now evaluate the possibility of constructing a large telescope with spherical mirrors intended for the observation of discrete cosmic sources of radiation. The estimates made in [1] show that to obtain the needed sensitivity the diameter of the main mirror of such a telescope must be on the order of 1500 mm. With such a large diameter it is necessary from size considerations to restrict the total length of the on-board instrument to ~2000 mm. It must be taken into account that the composition of the instrument must include a cryostat for cooling the radiation receiver, the height of which is ~1000 mm for a prolonged period of retention of the liquid helium. Thus, the length of the optical system of the instrument should not exceed ~1000 mm. Considering that under these conditions the entrance window of the receiver in the cryostat has to be located almost flush with the main mirror, we find that the working segment of the telescope must be very small, for example, 50 mm.

Based on the figures presented in Section 1 for the focal distance and field of view of a telescope one can now calculate several variants of a telescope objective with the following initial parameters: $\Delta \simeq 1000$ mm, $F \simeq 3000$–5000 mm, $D = 1500$ mm, $2y' = 30$ mm, $2\theta = 20$–$30'$. The results of the calculations for five variants are summarized in Table 2.

Variants 1 and 3 must be discarded immediately since the corresponding values of $2\rho_m$ considerably exceed the diameter $2y'$ of the entrance window. Variant 4 is disadvantageous because of the small elongation ($\gamma \simeq 0.31$). In this variant the field of view could be reduced

TABLE 2. Variants of Calculation and
Correction of a Large Telescope Objective
with Spherical Mirrors

Parameters	Number of variant				
	1	2	3	4	5
Δ	800	900		1000	
F	4231	3500	7092	3248	5250
F_1	1116	1317	1180	1538	1355
γ	0.186	0.257	0.127	0.309	0.190
α	3.53	3.16	4.21	2.86	3.81
β	0.264	0.376	0.166	0.474	0.255
F_2	−430	−669	−336	−1023	−480
D_2	425	475	356	525	394
$2\rho_m$	42	24	54	14	32
x	349	158	737	87	328
$2\theta'$	24	29	14	32	20

Initial data: $D = 1500$, $2y' = 00$, $\bar{\rho}_m = 50$.

to 16' by decreasing the entrance window to 2y' = 15 mm. However, this would lead to the
necessity either of diaphragming the entrance window and thereby exposing the receiver to the
parasitic radiation of the diaphragm or of decreasing the diameter of the light pipe which
would be expressed in an increase in losses to reflection in it. There remain variants 2 and 5
with values of $2\rho_m$ close to 2y' and fields of view of 30 and 20', respectively. Both these variants satisfy our conditions. Variant 2 is simpler in realization thanks to the smaller relative
aperture of the mirror, although variant 5 is preferable from considerations connected with the
field of view. The final choice of the variant can be made only under concrete conditions based
on overall size, simplicity of realization, etc.

In the conclusion of this section we estimate the coma of a telescope objective with
spherical mirrors. It can be shown by the usual methods (see [3], for example) that for this
system the coma coefficient in the first approximation is equal to

$$K_{II} \simeq -1/4\beta^2.$$

Since the angular radius of the coma spot is

$$\theta_{II} = K_{II}A^2w/4, \tag{28}$$

where A is the relative aperture of the system and w is the angle between the axis of the beam
and the axis of the system, by substituting A = βA_1 and K_{II} from (28) here and exchanging w for
the half-angle θ of the field of view we find that

$$\frac{\theta_{II}}{\theta} \simeq \frac{A_1^2}{16}. \tag{29}$$

Thus, the coma of a telescope objective with spherical mirrors in the first approximation
depends only on the relative aperture of the main mirror. For A_1 = 1 : 1 the widening θ_{II} of the
image by the coma at the edge of the field of view (w = θ) is 16 times less than θ. Therefore
in our problem one can practically always neglect the widening due to the coma.

4. CASSEGRAIN REFLECTING TELESCOPE OBJECTIVE

WITH AN IMPERFECT PARABOLIC MIRROR

OF LARGE DIAMETER

In the preceding section we explored the possibility of constructing a large telescope
objective with a diameter of 1500 mm and a length of ~1000 mm from spherical mirrors.
However, the fabrication and testing of large mirrors with a relative aperture of ~1 : 1 are
connected with great difficulties even when their shape is spherical. As far as we know there
are no finished mirrors with such parameters. Therefore here we will consider the possibility
of implementing a large Cassegrain telescope objective in its classic variant (paraboloid +
hyperboloid), adopting as the main mirror a ready-made searchlight mirror of paraboloidal
shape 1500 mm in diameter.

As mentioned in Section 1, in such a variant the spherical aberration of each mirror and
of the system as a whole proves to be corrected, at least with the accuracy of higher-order
terms. In adopting a searchlight mirror we come up against the fact that its parabolic shape
is imperfect, i.e., is subject to more or less random local deviations on the average over the
mirror. This is connected with the fact that the light source in the searchlight has finite
dimensions and there is no sense in striving for a very good mirror quality and giving it a focal
spot smaller than the size of the light source.

Thus, having escaped spherical aberration, we have encountered broadening of the image
of an infinitely distant point because of the insufficiently exact shape of the surface of the main

mirror. It must be studied how this fact influences the size of the focal spot of the telescope objective and the efficiency in using the energy of the focal spot at the entrance window of the receiver. We will assume that with its moderate diameter the hyperbolic secondary mirror can be fabricated exactly enough that its quality does not affect the image.

The parabolic searchlight mirrors which we have at our disposal have a focal distance of 640 mm at a light diameter of 1500 mm (relative aperture 1 : 0.425) and with an infinitely distant point source give a spot with a diameter of $2r \approx 5$ mm at the focus. Thus, the initial data for a paraxial calculation of the telescope objective are as follows: $D = 1500$ mm, $F_1 = 640$ mm, $2y' = 30$ mm, $2\theta = 20$-$30'$, $p = 50$ mm, $F = 3000$-5000 mm. Such a fast mirror allows one to achieve a shorter system for the same focal distance than the systems of Table 2. The results of the calculation of both variants (with $F = 3000$ and 5000 mm) are given in Table 3 (the dimensional values are in millimeters). The transverse magnification of the secondary mirror $V = 1/\beta$ and the eccentricity of the secondary mirror e_2^2 are presented here in addition to the values already described. The value V plays an important role: It shows how many times the spot of dispersion at the focus of the system is magnified in comparison with the focus of the main mirror (for a perfect secondary mirror). The diameter of this spot $2r' = 2rV$ is given in the last row of the table for both variants. In the variant with longer focus (and more preferable in terms of the geometrical field of view) the value of $2r'$ exceeds the diameter $2y'$ of the entrance window. Therefore the question arises: How efficient is the use of the energy of the focal spot at the entrance window of the receiver? In order to answer this question and to validly choose one of the two variants of the telescope objective it is necessary to study the energy distribution over the focal spot. This is necessary because, in contrast to the case of spherical aberration where the spot of least dispersion (with a mirror of perfect shape) has a more or less sharp boundary, the focal spot from a mirror of imperfect shape has practically no boundary and the intensity in it falls to zero only at infinity.

In order to simplify the computations we now change from the focal plane of the telescope objective to the focal plane of the main mirror and we will assume that the receiver entrance window with a diameter $2\rho = 2y'/V$ (the image of the window in the secondary mirror) is located in it.

When a plane wave from an infinitely distant point source falls on a mirror with a focal distance F and a mount radius R a perfect mirror produces at its focus the well-known diffraction image of the source with an angular radius $\alpha_D = 1.22\lambda/2R$ of the central spot. If the surface of the mirror is imperfect then a blurred spot of diffusion appears at the focus instead of a diffraction pattern (see [6], for example). With small deviations (compared with the wavelength λ) in the shape of the mirror from the perfect shape the energy concentrated in this spot is small compared with the energy in the diffraction pattern. With an increase in the mirror defects (or with a decrease in λ) the energy from the diffraction pattern passes into the spot of diffusion; with considerable defects almost all the energy is distributed over the broad spot of diffusion and the diffraction pattern is absent.

TABLE 3. Variants of a Large Cassegrain Telescope Objective
with Aspherical Mirrors

Parameters	$F = 3000$	$F = 5000$	Parameters	$F = 3000$	$F = 5000$	Parameters	$F = 3000$	$F = 5000$
β	0.214	0.128	F_2	−154	−90	V	4.7	7.8
α	5.26	8.18	D_2	285	184	$2\theta'$	34	21
Δ	518	562	e_2^2	2.37	1.78	$2r'$	24	39

Initial data: $D = 1500$, $F_1 = 640$, $2y' = 30$.

Let us make simplifying assumptions in order to find the intensity distribution in the focal spot. One of them we already introduced earlier: We will assume that the secondary mirror is perfect and we will study the focal spot of the main mirror. Second, we exclude consideration of systematic zonal errors in the main mirror which lead to defocusing or lateral displacements of the spot.

Thus, we will consider a concave mirror with a focal distance F, with the entrance and exit pupils coinciding with its mount radius R. The intensity distribution in the focal plane is found from the known distribution of amplitude and phase in the exit pupil using the Debye integral (see [7], for example). We will consider the deviation $\Delta(y, z)$ of the mirror surface from a perfect surface [6] as a random value with a zero average value $\Delta = 0$ (y and z are the Cartesian coordinates in the plane of the pupil). Then the deviation of the phase in the pupil from the perfect spherical wave (the Gauss comparison sphere) is $S(y, z) = 4\pi\Delta(y, z)/\lambda$, where $\bar{S} = 0$. The power spectrum of the fluctuations $S(y, z)$ is determined in accordance with the Wiener−Khinchin equation

$$P(k_y, k_z) = \int\int_{-\infty}^{\infty} K(\eta, \xi) \exp[2\pi i (k_y \eta + k_z \xi)] \, d\eta \, d\xi, \tag{30}$$

where k_y and k_z are the wave numbers along the y and z axes; $K(\eta, \xi)$ is the correlation function. If $S(y, z)$ forms a uniform and isotropic field of fluctuations with a normal distribution then the correlation function is conveniently given in the form

$$K(\eta, \xi) \approx \exp\left(-\frac{\eta^2 + \xi^2}{l^2}\right), \tag{31}$$

where l is the correlation radius. It corresponds to the power spectrum

$$P(k_y, k_z) \approx \exp(-\pi^2 k^2 l^2), \tag{32}$$

where $k^2 = k_y^2 + k_z^2$. Two conditions must be imposed on the correlation radius: $l \gg \lambda$ (the mirror surface is considered as specular) and $l \ll R$ (defocusing and lateral displacements of the focal spot are ignored).

Using the Debye integral one can now obtain [6] the average intensity distribution in the focal plane for the diffracted and scattered (by the mirror defects) light. For a point at a distance r from the optical axis we have for the diffracted light

$$I_D(r) = \frac{\pi I_0 R^2}{\lambda^2} e^{-\bar{S}^2} \frac{4 J_1^2\left(\frac{2\pi R}{\lambda F} r\right)}{\left(\frac{2\pi R}{\lambda F} r\right)^2} \tag{33}$$

and for the scattered light

$$I_s^{(1)}(r) = \frac{\pi I_0 l^2}{\lambda^2 \bar{S}^2} (1 - e^{-\bar{S}^2}) \exp\left[-\left(\frac{\pi l r}{\lambda F \sqrt{\bar{S}^2}}\right)^2\right] \quad \text{for} \quad \bar{S}^2 \geqslant 1, \tag{34}$$

$$I_s^{(2)}(r) = \frac{\pi I_0 l^2}{\lambda^2} (1 - e^{-\bar{S}^2}) \exp\left[-\left(\frac{\pi l}{\lambda F} r\right)^2\right] \quad \text{for} \quad \bar{S}^2 < 1. \tag{35}$$

Here $I_0 = \pi R^2 A_0^2/F^2$ is the total intensity of the light emerging from the pupil (A_0 is the amplitude in the pupil); $J_1(z)$ is a Bessel function of the first kind; $\bar{S}^2$ is the rms deviation of the phase from the comparison sphere.

The width of the central spot of the diffraction pattern is $\alpha_D = r_D/F = 1.22\lambda/2R$, while the width of the spot of scattered light $\alpha_s^{(1)} = r_s^{(1)}/F = \lambda\sqrt{\bar{S}^2}/\pi l$ does not depend on λ for large deviations of the mirror surface $\sqrt{\overline{\Delta^2}} \gg \lambda/4\pi$ and is determined entirely by the size of the deviations; for small deviations $\sqrt{\overline{\Delta^2}} \ll \lambda/4\pi$, we have $\alpha_s^{(2)} = r_s^{(2)}/F = \lambda/\pi l$, and the scattered spot is very broad but has a relatively low intensity, so that the diffraction pattern predominates.

To calculate the efficiency of the use of the focal spot energy at the receiver entrance window of radius $\rho = y'/V$ it is ncessary to integrate the distributions (33)-(35). Since the distributions are centrally symmetrical,

$$E = \int_0^{2\pi} \int_0^{\rho} I(r)\, r\, dr\, d\varphi \sim \int_0^{\rho} I(r)\, r\, dr. \tag{36}$$

Having completed the integration and converting to the rms error of the mirror $\varepsilon = \sqrt{\overline{\Delta^2}}$, we obtain

$$E_D(\rho) = \pi R^2 A_0^2 \exp\left[-\left(\tfrac{4\pi\varepsilon}{\lambda}\right)^2\right]\left[1 - J_0^2\left(\tfrac{2\pi R}{\lambda F}\rho\right) - J_1^2\left(\tfrac{2\pi R}{\lambda F}\rho\right)\right], \tag{37}$$

$$E_s^{(1)}(\rho) = \pi R^2 A_0^2 \left\{1 - \exp\left[-\left(\tfrac{l\rho}{4\varepsilon F}\right)^2\right]\right\} \qquad (\varepsilon \geqslant \lambda/4\pi), \tag{38}$$

$$E_s^{(2)}(\rho) = \pi R^2 A_0^2 \left(\tfrac{4\pi\varepsilon}{\lambda}\right)^2 \left\{1 - \exp\left[-\left(\tfrac{\pi l}{\lambda F}\rho\right)^2\right]\right\} \qquad (\varepsilon \leqslant \lambda/4\pi). \tag{39}$$

As $\rho \to \infty$ we find the total intensities in the corresponding distributions:

$$E_D = \pi R^2 A_0^2 \exp\left[-\left(\tfrac{4\pi\varepsilon}{\lambda}\right)^2\right], \tag{40}$$

$$E_s^{(1)} = \pi R^2 A_0^2 \qquad (\varepsilon \geqslant \lambda/4\pi), \tag{41}$$

$$E_s^{(2)} = \pi R^2 A_0^2 \left(\tfrac{4\pi\varepsilon}{\lambda}\right)^2 \qquad (\varepsilon \leqslant \lambda/4\pi). \tag{42}$$

From this it is seen that when $\varepsilon > \lambda/4\pi$ all the energy goes into the scattering spot, the intensity and size of which do not depend on λ, while the intensity of the diffraction pattern is equal to zero. When $\varepsilon < \lambda/4\pi$ the energy is distributed between diffraction and scattering, with the ratio of intensities being

$$\frac{E_s^{(2)}}{E_D} = \left(\tfrac{4\pi\varepsilon}{\lambda}\right)^2 \exp\left(\tfrac{4\pi\varepsilon}{\lambda}\right)^2 \sim \tfrac{\varepsilon^2}{\lambda^2}. \tag{43}$$

In the limiting case $\varepsilon \ll \lambda/4\pi$ (or $\lambda \gg 4\pi\varepsilon$) all the energy is concentrated in the diffraction pattern.

Using these results one can obtain estimates of the rms errors of our mirrors from the sizes of the spots of dispersion in visible light. For $\lambda_V = 5 \cdot 10^{-5}$ mm the angular width of the diffraction spot with R = 750 mm is $\alpha_{DV} \simeq 4 \cdot 10^{-7}$. A spot of diffusion with a radius of 2.5 mm (searchlight mirror) at F = 640 mm has an angular size $\alpha_s \simeq 4.7 \cdot 10^{-3}$, so that $\alpha_s/\alpha_{DV} \simeq 1.2 \cdot 10^{-4}$, i.e., the focal spots of our mirrors in visible light correspond to the case $\varepsilon \gg \lambda/4\pi$. Therefore from the equation $\alpha_s^{(1)} = 4\varepsilon/l$ we find for the searchlight mirror $\varepsilon_S/l \simeq 10^{-3}$. Since $l \ll R$ it is reasonable to assume that $l \approx R/20 = 37.5$ mm, from which $\varepsilon_S \approx 40 \mu$.

Since the efficiency of the use of the focal spot energy at the opening ρ is

$$\eta_{in} = E(\rho)/\pi R^2 A_0^2,$$

from (37)-(39) and (40)-(42) we find

$$\eta_D = \exp\left[-\left(\tfrac{4\pi\varepsilon}{\lambda}\right)^2\right]\left[1 - J_0^2\left(\tfrac{2\pi R}{\lambda F}\rho\right) - J_1^2\left(\tfrac{2\pi R}{\lambda F}\rho\right)\right], \tag{44}$$

$$\eta_s^{(1)} = 1 - \exp\left[-\left(\frac{l\rho}{4\varepsilon F}\right)^2\right] \quad (\lambda \leqslant 4\pi\varepsilon), \tag{45}$$

$$\eta_s^{(2)} = \left(\frac{4\pi\varepsilon}{\lambda}\right)^2 \left\{1 - \exp\left[-\left(\frac{\pi l}{\lambda F}\right)^2\right]\right\} \quad (\lambda \geqslant 4\pi\varepsilon). \tag{46}$$

The corresponding graphs are presented in Fig. 6 (F = 5000 mm, V = 7.5). The total efficiency η_{in} of the use of the focal spot energy as a function of the wavelength λ is obtained by summing all three graphs.

When $\lambda < 4\pi\varepsilon$ the efficiency is determined entirely by scattering and does not depend on λ (Fig. 6). When $\lambda \approx 4\pi\varepsilon$ the role of scattering begins to decrease rapidly while at the same time the diffraction spot is still small compared with the entrance window, and therefore the efficiency increases with an increase in λ. The increase in efficiency ceases when the diffraction spot is comparable to the entrance window. The absolute values of the efficiency η_{in} in the variant F = 5000 mm for $\lambda < 4\pi\varepsilon$ are relatively low: $\eta_{in} \approx 0.5$; η_{in} is clearly higher in the variant F = 3500 mm. However, the longer-focus variant is preferable for the following reasons.

The fields of view 2θ indicated above for the two variants are geometrical, i.e., they are determined from the radius of the receiver entrance window and the focal distance on the assumption that the focal spot has a small size in comparison with the window radius. In actual fact the focal spot has a size on the order of the window radius (even when $\lambda > 0.5$-0.75 mm, where diffraction rather than scattering predominates). Therefore when the point source is shifted from the optical axis of the telescope by an angular distance θ, i.e., when the spot in the focal plane is displaced by a distance equal to the radius of the entrance window, part of the spot's energy will fall on the receiver. Thus, the effective field of view θ_e, i.e., the angular distance of a point source from the optical axis at which power from the source still enters the receiver, will be greater than the geometrical field of view θ. To be definite, we arbitrarily consider the effective field of view θ_e to be that angular distance between the point source and the optical axis at which the power entering the receiver decreases by e times compared with the power for the source in the axial position.

To determine θ_e it is necessary first of all to find the dependence of the power E falling on the entrance window on the angular distance w between the source and the optical axis. In this case the broadening of the spot due to the coma can be ignored. Actually, the angular broadening of the spot caused by the coma at an angular displacement w is determined by Eq. (28). If we set the coma coefficient K_{Π} equal to 1/4 for a parabolic mirror (and for a telescope objective containing a parabolic main mirror) and A = D/F (D = 1500 mm and F = 5000 mm) we find that $\theta_{\Pi}/\theta < 1/100$ for w = θ. Thus, the angular broadening of the focal spot due to the coma is negligibly small.

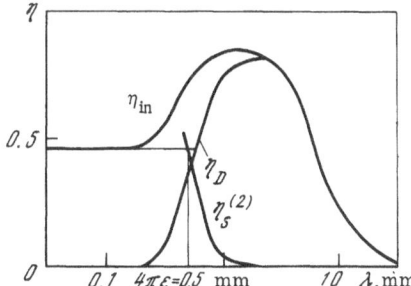

Fig. 6. Efficiency η_{in} of use of focal spot energy at entrance window of receiver for a telescope objective with a parabolic searchlight mirror.

In order to find the dependence of E on w one must integrate the intensity distribution in the spot over the area of the entrance window for different distances w between the source and the axis, i.e., for different displacements r_0 between the axis of the light pipe and the axis of symmetry of the distribution (Fig. 7).

Let us introduce in the focal plane of the optical system (Fig. 7) the polar coordinates r, φ with the pole at the center of symmetry of the focal spot. It is clear that with an intensity distribution $I(r, \varphi)$ in the spot the power falling on the entrance window is proportional to the integral of $I(r, \varphi)$ over the entrance window σ:

$$E \approx \iint\limits_{(\sigma)} I(r, \varphi)\, d\sigma. \tag{47}$$

If the distance between the axis of the light pipe and the pole is designated as r_0 and the angle φ is measured from the direction of r_0 then the equation of the circle bounding the window when $r_0 \geq \rho$ will be

$$r = \begin{cases} r_0 \cos \varphi - \sqrt{\rho^2 - r_0^2 \sin^2 \varphi} = r_1(\varphi), \\ r_0 \cos \varphi + \sqrt{\rho^2 - r_0^2 \sin^2 \varphi} = r_2(\varphi) \end{cases} \tag{48}$$

and the integral (47) takes the form ($d\sigma = r\,dr\,d\varphi$)

$$E(r_0) \approx \int\limits_{-\varphi_m}^{\varphi_m} d\varphi \int\limits_{r_1(\varphi)}^{r_2(\varphi)} I(r, \varphi)\, r\,dr, \tag{49}$$

where $\varphi_m = \arcsin(\rho/r_0)$ is the direction of the radius vector r_m tangent to the circle of the window, while $r_1(\varphi)$ and $r_2(\varphi)$ are the equations of the inner and outer arcs of the circle (48). When $r_0 < \rho$ the limits of integration are changed:

$$E(r_0) \approx \int\limits_{0}^{2\pi} d\varphi \int\limits_{0}^{r_2(\varphi)} I(r, \varphi)\, r\,dr. \tag{50}$$

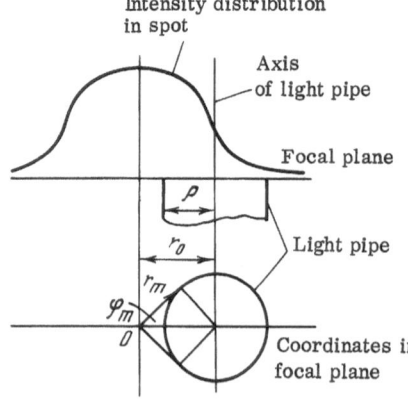

Fig. 7. For calculation of the effective field of view.

Let us find the expression for E in the two extreme cases: pure scattering ($\lambda \lesssim 100\ \mu$) and pure diffraction ($\lambda \gtrsim 1$ mm). The appropriate equations for I have the form

$$I_s^{(1)} \approx \exp[-(br)^2], \tag{51}$$

where b = $(l/\varepsilon)/4F$, and

$$I_D \approx \frac{J_1^2(ar)}{(ar)^2}, \tag{52}$$

where $a = 2\pi R/\lambda F$. Substituting these expressions into Eqs. (49) and (50) and integrating over r, we obtain

$$E_s^{(1)}(r_0) \approx \begin{cases} \dfrac{1}{2b^2}\left(2\pi - \int\limits_0^{2\pi} \exp[-b^2 r_2^2(\varphi)]\,d\varphi\right) & (r_0 < \rho), \\[4mm] \dfrac{1}{b^2}\int\limits_0^{\varphi_m} (\exp[-b^2 r_1^2(\varphi)] - \exp[-b^2 r_2^2(\varphi)])\,d\varphi & (r_0 \geqslant \rho); \end{cases} \tag{53}$$

$$E_D(r_0) \approx \begin{cases} \dfrac{1}{2a^2}\left(2\pi - \int\limits_0^{2\pi} [J_0^2(ar_2(\varphi)) + J_1^2(ar_2(\varphi))]\,d\varphi\right) & (r_0 < \rho), \\[4mm] \dfrac{1}{a^2}\int\limits_0^{\varphi_m} [J_0^2(ar_1(\varphi)) + J_1^2(ar_1(\varphi)) - J_0^2(ar_2(\varphi)) - J_1^2(ar_2(\varphi))]\,d\varphi & (r_0 \geqslant \rho). \end{cases} \tag{54}$$

The normalized dependence of the telescope sensitivity on the direction toward the source (the directional diagram) is determined by the ratio $\psi(r_0) = E(r_0)/E(0)$. The integrals (53) and (54) entering into the expressions for $\psi_s^{(1)} = E_s^{(1)}(r_0)/E_s^{(1)}(0)$ and $\psi_D = E_D(r_0)/E_D(0)$ in the general case of an arbitrary displacement r_0 do not have exact analytical expressions. Therefore to obtain the directional diagrams ψ_s and ψ_D we undertook numerical integration in (53) and (54) using a computer for the variant of a telescope with a searchlight mirror (R = 750 mm, F = 640 mm, $l/\varepsilon = 10^3$, $\rho = 2$ mm, a = 7.364, b = 0.39).

The results of the calculations are presented in Fig. 8. At the wavelengths ($\lambda > 0.5$ mm) where the effect of the imperfect nature of the mirror is absent the directional diagram has the characteristic diffraction form (for $\lambda = 1$ mm the radius ρ of the image of the receiver entrance window is about four times greater than the radius of the central spot of the diffraction pattern). The sensitivity of the instrument in this extreme case falls to ~1/2, the maximum level at $r_0 \simeq \rho$, i.e., the effective field of view θ_{eD} hardly differs from the geometrical field of view.

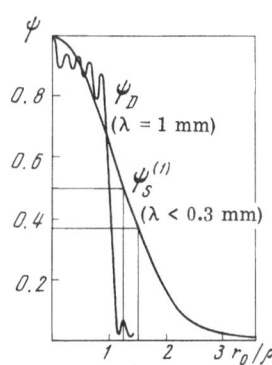

Fig. 8. Normalized dependence of telescope sensitivity on direction toward the source (directional diagram) for the cases of pure diffraction (ψ_D) and of pure scattering ($\psi_s^{(1)}$).

At shorter wavelengths (λ < 0.3 mm), where deviations of the mirror surface from a perfect surface play the dominant role, the directional diagram (for a normal distribution law for the mirror errors) takes on a different appearance, smoother with relatively flat slopes. Based on the sensitivity level of 1/e of the maximum level the effective field of view θ_{es} exceeds the geometrical field of view θ by 1.5 times. Since $\theta \simeq 20'$ for this variant, $\theta_{es} \simeq 30'$.

Thus, because of widening of the effective field of view we are induced to end up with the longer-focus variant of the telescope objective, having sacrificed somewhat the efficiency of the use of the focal spot energy at the entrance window of the receiver in the process.

5. OPTICAL SYSTEM FOR A LARGE

SUBMILLIMETER TELESCOPE

We have calculated a reflecting telescope objective, striving to obtain from an infinitely distant source a spot of the smallest possible size at the focus at a fixed distance (p = 50 mm) from the apex of the main mirror. Considering that the entrance window of the receiver lies at the focus of the system, we also calculated the efficiency of the use of the focal spot energy sizes of the focal spot and the window. Now it is necessary to introduce important elements into the system — the modulator, calibrator, and mechanism for changing filters — and to combine all this into a compatible optical system for the telescope. The complete optical system of the telescope is presented in Fig. 9. Before describing it let us dwell on the means of modulation of the radiation.

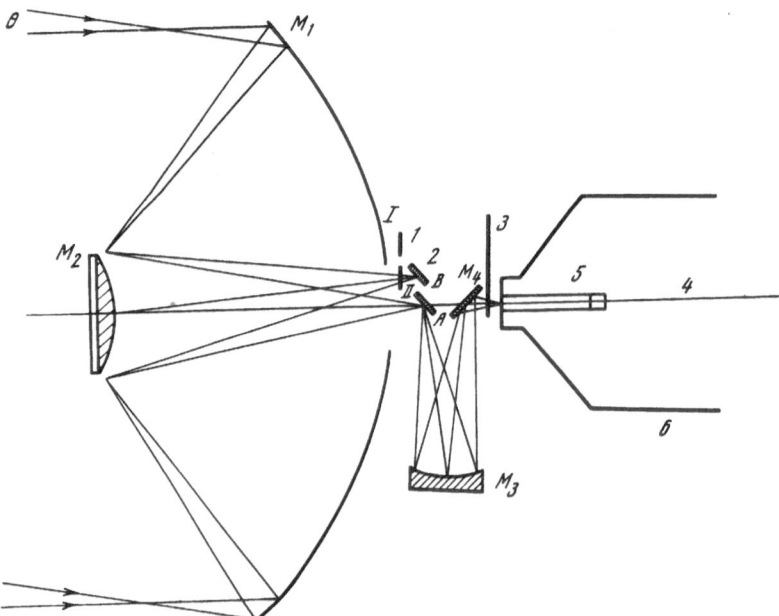

Fig. 9. Optical system of a large submillimeter telescope. 1) Calibrator; 2) modulator; 3) turret containing changeable filters; 4) optical axis of telescope; 5) receiver; 6) cryogenic system (cryostat).

We employed modulation of the useful signal in the submillimeter telescope, as in telescopes of the IR and SHF ranges, to exclude the effect on the telescope's threshold sensitivity
of anomalous receiver noises developing near the zero frequencies. Here the amplification
takes place at a modulation frequency selected from the consideration that the spectral density
of the radiometer noise should be close to its asymptotic value at high enough frequencies.
The modulation frequency is taken as 200 Hz in practice with allowance for constructional considerations (see below).

There are as yet no low-noise, efficient (with large modulation depth), nonmechanical
modulators in the submillimeter range of interest to us. Therefore the modulation is accomplished mechanically. Two methods of modulation are possible in principle: 1) amplitude modulation, in which the radiation arriving at the telescope from the source is replaced at the
receiver entrance by the radiation of an equivalent (in particular, by the opaque absorbing or
reflecting surface of the modulator itself, the shutter); and 2) "diagram" modulation, in which
periodic swinging of the telescope field of view occurs so that radiation from the source is
compared with the radiation of the section of the celestial sphere adjacent to it. Let us examine
these methods.

In the observation of rather faint discrete sources of radiation amplitude modulation of
the flux proves to be extremely disadvantageous. The main reason for this is that the power
of the radiation arriving from the source is compared with the power of the radiation supplied
from the modulator, which is at a temperature on the order of 250-300°K, while the voltage
measured at the amplifier output is proportional to the difference in these powers. In the submillimeter range the power of the radiation of bodies heated to room temperature ($\sim 300°K$) is
roughly proportional to the temperature (the Rayleigh–Jeans law). Since the modulator must
cover the entire field of view of the receiver, the power of the radiation coming from it (re-
gardless of the dependence on its emissivity) proves to be considerably greater than the power
of the radiation coming from the point source. As a result, in order not to exceed the allowable
limits of the output voltage it is necessary to decrease the gain and to measure a very small
signal from the source against the background of a large signal from the modulator.* Let us
clarify this with a calculation.

The telescope collects from a point source on the optical axis the power

$$P_s = \Phi_s C, \tag{55}$$

where C is the area of the mirror and Φ_s is the flux in the effective sensitivity band $\Delta \nu_e$ (all
subsequent arguments are conducted for this band without loss of generality). Of this power
$\eta_{in} P_s$ enters the receiver, where η_{in} is the efficiency at the light pipe. In observing the background near the source the telescope collects the power

$$P_b = B_b \Omega_e C, \tag{56}$$

where B_b is the background brightness and $\Omega_e = \pi \theta_e^2$ is the solid angle of the effective field of
view of the telescope. The modulator sends into the light pipe the power

$$P_m = B_m \omega a, \tag{57}$$

where B_m is the brightness of the modulator, $\omega = \pi \beta_0^2$ is the solid angle of the field of view of
the receiver, and a is the area of the entrance window. The main and secondary mirrors of

---

*We shall not consider the cases of deep cooling of the modulator or of automatic compensa-
tion of the temperature difference which are possible in principle but difficult to accomplish
in practice.

the telescope (visible from the light pipe over the same angle β_0) radiate into the light pipe the
power

$$P_t = B_t \omega a, \tag{58}$$

where B_t is the brightness of each mirror.

At the temperature $T \approx 300°K$ the brightness of the modulator and the mirror can be
taken as equal to

$$B \simeq 3.1 \cdot 10^{-28} \nu^2 \alpha T, \tag{59}$$

where α is the absorption coefficient. Henceforth we will assume that the modulator is spec-
ular and $\alpha_t \simeq \alpha_m \simeq 10^{-2}$. In this case $B \simeq 10^{-3}$ W/m$^2 \cdot$ sr for $T \approx 300°K$ and $\nu \simeq 10^{12}$ Hz
$(\lambda \simeq 100 \ \mu)$.

When observing the source the output voltage is

$$u_s = KS[P_m - (\eta_{in}P_s + P_b + P_t)] \tag{60}$$

and when observing the background

$$u_b = KS[P_m - (P_b + P_t)]. \tag{61}$$

Here K is the gain of the amplifier and S is the volt-watt sensitivity of the receiver. The rms
voltage fluctuations when receiver noise dominates are the same in both cases: $\delta u = KSP_N/\sqrt{\tau}$
(τ is the time constant and P_N is the minimum detectable power). The difference $P_m - P_t$ of
large values enters into both these equations, which it is difficult to maintain as equal, however, since first, the modulator and mirror are separated spatially, and second, P_m includes,
generally speaking, the power of the scattered radiation of telescope details reflecting from
the modulator into the light pipe (it is also $\sim P_m$). Therefore, from now on we will assume that
$P_m \gg P_t$. We can now show that $P_m \gg P_s$ and P_b. For $\beta_0 = 8°40' \simeq 0.15$ rad we have $\omega =$
$2.25 \cdot 10^{-2}$ sr; since the diameter of the entrance window is 30 mm, $a = 7 \cdot 10^{-4}$ m^2 and $\omega a =$
$5 \cdot 10^{-5}$ m$^2 \cdot$ sr. Then $P_m \simeq B_m \omega a = 5 \cdot 10^{-8}$ W. At the same time, for a source with a flux
density $F_s \approx 10^{-23}$ W/m$^2 \cdot$ Hz we have $P_s \approx 10^{-11}$ W, from which it is seen that $P_m \gg P_s$ even
with a specular modulator (and $P_m \gg P_b$ if the background is faint). In this case it is necessary
to have a gain of no more than

$$K_{max} = \frac{u_{max}}{SP_m}, \tag{62}$$

where u_{max} is the maximum allowable output voltage (the scale of the measuring instrument).
Then the measure of the signal of interest to us on the output scale is equal to

$$\frac{\Delta u}{u_{max}} = \frac{u_s - u_b \pm 2\delta u}{u_{max}} = \frac{u_s - u_b \pm 2\delta u}{K_{max}SP_m}.$$

Having made all the substitutions, we obtain

$$\frac{\Delta u}{u_{max}} = -\frac{\eta_{in}P_s}{P_m} \pm 2\frac{P_N\sqrt{\tau}}{P_m}, \tag{63}$$

i.e., the useful signal comprises an insignificant fraction of the output scale which is entirely
filled by the signal from the modulator.

All that has been said about amplitude modulation applies in equal measure to the observation of point sources against a powerful background (for example, a source setting behind
the earth's atmosphere); the background now exerts the same effect as the modulator. True,
in this case one can improve the source-to-background ratio by decreasing the field of view,

since

$$\frac{P_s}{P_b} = \frac{\Phi_s}{B_b \Omega_s},$$

which, as we have seen, is not always successful.

Diagram modulation, when the mirror modulator alternately directs the radiation of the source and the radiation of the background near the source into the receiver, proves to be more advantageous in the case of the measurement of discrete sources.

For a point source

$$u_s = KS\left[\eta_{in}P_s + (P_m - P'_m) + (P_t - P'_t)\right] = KS\left[\eta_{in}P_s + \delta P_m + \delta P_t\right],$$

where P_m and P_t are radiation powers for one modulator position and P'_m and P'_t are the same for the other modulator position. Similarly, when observing a section next to the source,

$$u_b = KS\left(\delta P_m + \delta P_t\right). \tag{64}$$

The voltage fluctuations as before are equal to $\delta u = KSP_N/\sqrt{\tau}$. Thus, the signal of interest to us is equal to

$$\Delta u = u_s - u_b \pm 2\delta u$$

with an accuracy of the unbalance of the powers δP_m and δP_t during modulation. Even if δP_m, $\delta P_t \approx \eta_{in}P_s$, diagram modulation is also more advantageous than amplitude modulation in this case, since the maximum voltage will be determined by the value $\delta P_m + \delta T_t + \eta_{in}P_s$, i.e., $\eta_{in}P_s$ will have a measure on the order of half the output scale.

Equation (64) is obtained on the assumption that the background is uniform, i.e., that $P_b = P'_b$. If the background is irregular then the term $\delta P_b = P_b - P'_b$ appears in (64). Its value is proportional to the amplitude of the background irregularity and is the larger, the closer the characteristic angular size of the irregularities is to the angle of swinging of the field of view during diagram modulation. To exclude δP_b it is necessary to choose the amplitude of the diagram modulation so that the angle of swinging of the field of view is small compared with the dimensions of the background irregularities. Since the irregularity of the background is different in the case of each specific source, we select the amplitude of the diagram modulation as minimal from constructional considerations (see below).

Let us estimate the balancing necessary in order that $\delta P_m \approx \eta_{in}P_s \approx 10^{-11}$ W. One can write δP through the difference δT in the temperatures of the modulator in the two positions:

$$\delta P \simeq 3.1 \cdot 10^{-28}\nu^2 a\omega a \delta T.$$

From this with the known numerical values of the parameters we find that $\delta P \approx 2 \cdot 10^{-10} \, \delta T$ [W], so that $\delta T \approx 0.1°$K for $\delta P \approx 10^{-11}$ W. This shows that with diagram modulation rather exact balancing of the temperatures of the modulator and the mirrors is required in order to realize the use of the scale indicated above. All these estimates are made for the frequency $\nu \approx 10^{12}$ Hz ($\lambda \approx 100\ \mu$). The requirements will be less rigid at longer wavelengths ($\nu \approx 10^{11}$ Hz, $\lambda \approx 1$ mm).

If the values of δP_m and δP_t (as can be assumed) vary little during the passage of the source through the field of view of the telescope, then the demands on the accuracy of balancing are reduced and are determined only by considerations of the efficiency of use of the scale. In the choice of the construction of the diagram modulator we rejected the swinging mirror because of constructional difficulties and settled on the system described below.

The diagram modulator is located in the focal plane (Fig. 9) and consists of two plane pivoted mirrors. One of them consists of two sectors; A is movable and pivots about an axis perpendicular to its plane and outside the boundary of the mirror.* Thus, sector A of the mirror is periodically (with a frequency of 200 Hz) inserted into the optical axis and cuts off the continuous stationary mirror B which is displaced from the axis. Thus the modulator alternately reflects the flux from the source, which is on the optical axis, and from the background next to the source. To obtain sufficiently deep modulation the angular distance between the beams incident on the receiver from mirror A and B of the modulator should not be less than the effective field of view θ_e. For F = 5000 mm (θ_e = 30') the distance between mirrors A and B should be no less than 45 mm.

In the problem of measuring the power of the background radiation diagram modulation is obviously unsuitable (it allows one to obtain signals proportional to the difference in brightnesses of adjacent sections). In this case amplitude modulation is needed. The radiation from some local emitter must serve for comparison with the radiation coming from the telescope. A calibrator-standard inserted in front of mirror B can play the role of such an emitter.

The calibrator (Fig. 9) consists of a plate with a known radiation brightness. It is normally located in position I; during calibration (and in the mode of amplitude modulation) the plate is inserted into the reference beam (position II) and cuts it off. In this case the power of the source radiation is compared not with the power of the background radiation but with the known power of the calibrator radiation. Such a calibrator represents a secondary power standard for monitoring the constancy of the telescope sensitivity during operation. Absolute calibration must be conducted against discrete sources with known brightness (such as Mars and Jupiter, see [1]).

With such a modulator construction it is impossible to locate the cryostat containing the receivers so that the entrance to the light pipe lies at the focus of the system. Therefore the spherical projection mirror M_3, which transfers the image from the focus of the telescope to the entrance window, is introduced into the telescope. The placing of the modulator at the focus is also dictated by the choice of the working segment p = 50 mm, which was discussed in Section 3. With F = 5000 mm mirror M_3 must have a focal distance F_3 = 250 mm and a diameter D_3 = 150 mm. The angles between the beams incident on the mirror and reflected from it are 10-15°, so that the coma and astigmatism of M_3 do not introduce serious broadening into the image of the focal spot.

The plane mirror M_4 deflects the beam into the entrance of the light pipe, near which is located a turret containing changeable filters for the isolation of individual subranges.

Finally, let us estimate the total efficiency of the telescope, allowing for all the losses between the main mirror and the receiver. As we saw in Section 4, the efficiency of the use of the focal spot energy at the entrance to the light pipe for the wavelength $\lambda \approx 100 \mu$ is $\eta_{in}(100) \approx 0.5$, and for $\lambda \approx 1$ mm it is $\eta_{in}(1) \approx 0.8$. According to estimates made in [2], the efficiency of the receiver itself is $\eta_r \approx 0.2$-0.3. The losses during reflection from the telescope mirrors can be considered as negligibly small since aluminum has a reflectivity of ~0.99 in the submillimeter range. Thus, the total efficiency of the telescope is approximately equal to $\eta \simeq \eta_{in} \eta_r$ and is $\eta(100) \approx 0.12$-0.15 for $\lambda \approx 100 \mu$ and $\eta(1) \approx 0.2$-0.25 for $\lambda \approx 1$ mm.

I wish to thank A. E. Salomonovich for constant interest in the work and valuable discussions.

---

* The modulator can be made to turn by a hermetically sealed motor using a magnetic clutch of the type developed for the Obzor instrument (see [8]).

LITERATURE CITED

1. A. E. Salomonovich and A. S. Khaikin, Tr. FIAN, 77, 33 (1974).
2. A. A. Kobzev, V. I. Lapshin, S. V. Solomonov, and A. S. Khaikin, Tr. FIAN, 77, 80 (1974).
3. D. D. Maksutov, Astronomical Optics [in Russian], Gostekhizdat, Moscow (1946).
4. J. E. Beckman and J. A. Shaw, Infrared Phys., 12, 219 (1972).
5. A. A. Kobzev, V. I. Lapshin, V. F. Troitskii, and A. S. Khaikin, Tr. FIAN, 77, 110 (1974).
6. H. Scheffler, Z. Astrophys., 55, 1 (1962).
7. M. Born and E. Wolf, Fundamentals of Optics [Russian translation], Mir, Moscow (1970), Ch. 8.
8. A. E. Salomonovich, S. V. Solomonov, A. S. Khaikin, and V. N. Gusev, Preprint FIAN, No. 126 (1974).

A TWO-CHANNEL COOLED RECEIVER FOR ON-BOARD
TELESCOPES OF THE SUBMILLIMETER RANGE

A. A. Kobzev, V. I. Lapshin, S. V. Solomonov,
and A. S. Khaikin

The possibility of studying the radiation of discrete astronomical sources in the submillimeter range of the spectrum using an on-board telescope is determined to a considerable extent by the threshold sensitivity of the instrument, which depends entirely on the size of the mirror and the threshold sensitivity of the radiation receiver. Estimates of the expected fluxes from sources [1] and of the efficiencies of telescope optical systems [2] show that with a mirror diameter of 1.5 m the radiation receiver must have a threshold sensitivity no worse than 10^{-11} W/Hz$^{1/2}$ at a width of 1000 GHz for the region of sensitivity in order to reliably record the radiation of a number of interesting sources [1]. To cover the greatest possible part of the submillimeter range with one instrument it is desirable to use a wide-band receiver. The necessity of mounting the instrument on board a flying vehicle imposes a number of specific requirements: relatively small overall size, weight, and energy consumption and stability against vibrations and accelerations.

Only semiconducting photoresistors cooled to a temperature of ~4°K satisfy this complex of conditions most completely: n-InSb [3-5] with sensitivity in the region of $\lambda > 300~\mu$, n-GaAs [6] sensitive in the region of $\lambda \approx 300~\mu$, and Ge containing different doping admixtures [6-8], of which a boron admixture gives a photoresistor sensitive from ~60 to ~140 μ. The threshold sensitivities of these photoresistors are of the same magnitude ($\delta P \approx 10^{-12}$ W/Hz$^{1/2}$), while their reception regions complement one another. Therefore, to create a receiver* which covers the greatest possible part of the submillimeter range it is necessary to use several different photoresistors at once. A receiver which combines receiving elements of n-InSb and Ge:B as the most highly developed in a technological respect is described here. Questions of the construction of an on-board cryogenic system for cooling the receiver represent a separate problem and are considered in [9].

In the development of the receiver we started from the following considerations. First, for elimination of parasitic exposure of the receiving elements to the radiation of "warm" surrounding objects and consequently to reduce the noise of the receiving elements the receiver must have a field of view compatible with the exit aperture angle of the telescope objective. Second, the size of the entrance window of the receiver should match the size of the focal spot

---

* In future we will call the optical device containing the receiving elements the photoresistors and the apparatus for cooling them with liquid helium, the receiver.

of the telescope objective. Third, power losses between the entrance window and the receiving elements must be reduced to a minimum. Finally, the two receiving elements must operate simultaneously not hindering one another, and have identical fields of view. In addition, the necessity of cooling imposes on the receiver construction the requirement of the minimum additional inflow of heat to the coolant (liquid helium).

A receiver consisting of a light pipe, a cone, and a spherical integrating chamber with the two receiving elements in it (Fig. 1) was constructed and prepared in accordance with these considerations. Such a construction is preferable to the direct (without using a light pipe) irradiation of the receiving elements through transparent windows in the cryostat, since it not only improves the reliability of the cryostat (thanks to the absence of "cold" windows) but also increases the duration of liquid helium retention. Let us examine briefly the separate parts of the receiver.

The inner diameter of the cylindrical light pipe (30 mm) satisfies the contradictory requirements of low inflow of heat to the liquid helium and small radiation power losses. In addition, it conforms well with the size of the focal spot of a telescope with a 1.5-meter mirror and a focal distance of 5 m [2], for which the receiver is primarily intended. The wall thickness of the light pipe (0.2 mm), the length together with the cone (375 mm), and the material (stainless steel) were chosen so as to obtain the lowest possible heat conduction without impairing the durability of the structure. Estimates of the losses based on the equations of [10] show that no more than 20% of the radiation power is lost in such a light pipe with a coefficient of absorption [11] in the walls $\alpha \lesssim 10^{-2}$ and an aperture angle [2] of the field of view $\theta_r \simeq 9°$.

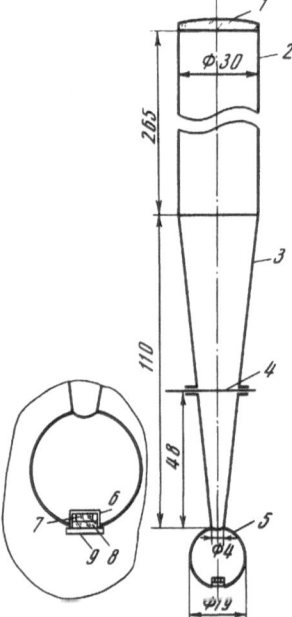

Fig. 1. Constructional diagram of receiver. 1) Collecting lens; 2) light pipe (0.2 mm stainless steel); 3) cone (0.2 mm stainless steel); 4) pressurizing film (polyethylene terephthalate) 20 μ thick; 5) integrating chamber; 6) cup (0.5 mm Teflon); 7) Ge:B photoresistor; 8) n-InSb photoresistor; 9) holder.

Between the light pipe and the integrating chamber is located a conical transition which increases the aperture angle of the beam from $\theta_r \simeq 9°$ to $\theta_c \simeq \pi/2$ and forms the field of view of the receiver. With the indicated θ_r, θ_c, and diameter D = 30 mm of the entrance opening of the cone the value d = 4.5 mm is obtained from the Lagrange—Helmholtz equation for the diameter of the exit opening. However, with these parameters D, d, and θ_r the length of the cone which provides the fullest possible suppression of parasitic exposure through the edge of its entrance opening proves to be excessively great. Therefore we used a cone of moderate dimensions in combination with a collecting lens [12], which permits one to almost completely avoid parasitic exposure and to receive only the radiation coming from the exit pupil of the telescope objective.

The length of the cone was chosen as 110 mm so as to provide a critical angle θ_r for the center of its entrance opening. With a distance of ~500 mm to the exit pupil (as in the variant of the 1.5-m telescope [2]) the thickness of a lens of quartz crystal which provides an image of the center of the pupil at the apex of the cone is small (2-2.5 mm) so that the losses in it do not exceed ~20%. To simplify the construction the lens is placed at the "warm" end of the light pipe, which is allowable if the light pipe does not alter the convergence of the beams. The lens is sealed and serves as the entrance window of the receiver, hermetically sealing the cavity of the light pipe.

The integrating chamber [12-14] of spherical shape containing the two receiving elements is located at the exit opening of the cone. An experiment is known [15] in which such a chamber provided for the parallel operation of four receiving elements.

In application to the radiation receiver the integrating chamber must be characterized by its efficiency $\eta_r = \Phi_D/\Phi$, i.e., by the fraction of power Φ_D absorbed in the receiving element compared with the power Φ supplied by the cone to the entrance opening of the chamber. Adopting the approach developed in [16], for a spherical chamber with diffusely reflecting walls (reflection coefficient $r \simeq 1$) we can find that, independent of the size of the chamber,

$$\eta_r = \frac{(1-r_D)\,A_D/A_0}{1+(1-r_D)\,A_D/A_0},\tag{1}$$

where A_D and A_0 are the areas of the receiving element and of the entrance opening, respectively, and r_D is the reflection coefficient of the receiving element. This expression is obtained for a single receiving element in the chamber; with elements of not very large size, however, they can be considered independently if they affect each other weakly (as, for example, in the case of n-InSb and Ge:B, each of which is transparent in the region of spectral sensitivity of the other).

The value η_r must be compared with the efficiency η_{r0} of a receiving element placed directly at the exit opening of the cone. For the ratio η_r/η_{r0} we obtain

$$\frac{\eta_r}{\eta_{r0}} = \begin{cases} \dfrac{A_D/A_0}{1+(1-r_D)\,A_D/A_0} & \text{for } A_D \geqslant A_0, \\[2ex] \dfrac{1}{1+(1-r_D)\,A_D/A_0} & \text{for } A_D < A_0. \end{cases}\tag{2}$$

Graphs of Eqs. (1) and (2) are shown in Fig. 2. As is seen, when $A_D \approx A_0$ the receiving element in the chamber always has a somewhat worse efficiency than in direct placement behind the opening of the cone. Despite this, for the combining of two or several receiving elements the integrating chamber still proves preferable since it provides fully identical fields of view for all the elements and moreover considerably simplifies the problem of locating and mounting the elements.

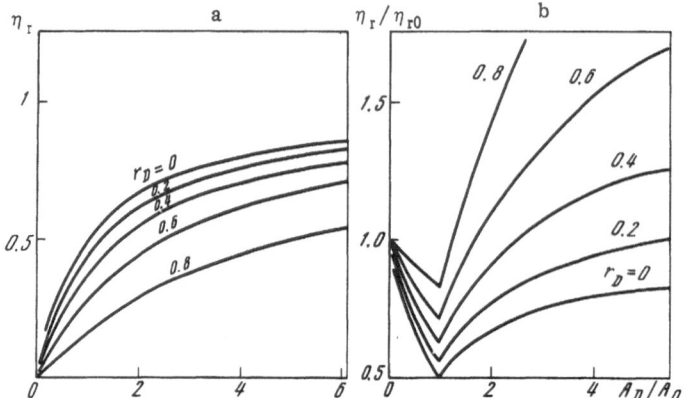

Fig. 2. Efficiencies of receiving element in chamber as functions of
A_D/A_0 for different values of r_D (a); ratios of efficiencies of receiving element in chamber and with direct mounting at opening of cone (b).

We used receiving elements of n-InSb $5 \times 5 \times 1$ mm in size and of Ge:B $3 \times 3 \times 1.5$ mm
in size. The reflection coefficients of these materials are about the same $(r_D \simeq 0.4)$. The
diameter of the chamber is limited by the free space in the neck of the cryostat and equals
19 mm; the diameter of the entrance opening is 4 mm. The chamber is made of copper and is
silver-plated; to obtain diffuse reflection the walls of the chamber have a roughness of ~100-
500 μ. From the sizes of the receiving elements and of the entrance opening one can find using
the curves of Fig. 2 that $\eta_r \simeq 0.5$ and $\eta_r/\eta_{r0} \simeq 0.9$ for n-InSb while $\eta_r \simeq 0.3$ and $\eta_r/\eta_{r0} \simeq 0.7$
for Ge:B. Thus, while sacrificing ~10-30% efficiency, we are assured of more uniform condi-
tions of operation and a simpler arrangement of the two receiving elements.

For estimates of the receiver sensitivities for both channels (n-InSb and Ge:B) we as-
sume that the total receiver efficiency for one of the channels is equal to $\eta = \eta_l \eta_p \eta_r$, where
η_l and η_p are the efficiencies of the lens and the light pipe, which are ~0.8 and ~0.7, respectively. Inserting here the values of η_r found above we obtain $\eta_1 \simeq 0.3$ (n-InSb) and $\eta_2 \simeq 0.2$
(Ge:B). From this we find that the threshold powers converted to the entrance window must be
~3 and ~5 times worse than the threshold powers of the n-InSb and Ge:B receiving elements,
respectively, because of losses in the channel.

The short-wave radiation is absorbed in a filter of black polyethylene in addition to the
collecting lens. This filter is attached to a polyethylene terephthalate film 20 μ thick which
is held between flanges of the cone (Fig. 1) using a ring seal of indium. The film serves to
seal off the cavity of the light pipe between the film and the lens. The cavity is evacuated to
prevent the possible convection of gaseous helium and the excessive inflow of heat to the liquid
helium connected with it. The region of spectral sensitivity of the receiver is from ~60 to
~140 μ (Ge:B) and from ~300 μ to ~30 mm (n-InSb). The long-wave limit of sensitivity is determined by diffraction at the exit opening of the cone.

Measurements of the field of view and sensitivity of the receiver were made with the
receiving elements placed near each other on the walls of the chamber opposite the entrance
opening. The field of view was measured using a source of small angular size; the true value
of the field of view coincides with the measured value $(\theta_r = 9 \pm 1°)$ with the accuracy of the
measurement errors. The sensitivity was measured for two black sources completely cover-

ing the receiver field of view and having brightness temperatures of 290 and 370°K. The threshold powers at the receiver entrance window have the order of $\delta P_1 \simeq 3 \cdot 10^{-12}$ W/Hz$^{1/2}$ (n-InSb) and $\delta P_2 \simeq 3 \cdot 10^{-11}$ W/Hz$^{1/2}$ (Ge:B). Taking the values of η_{r1} and η_{r2} into account we obtain $\delta P_{01} \simeq 10^{-12}$ W/Hz$^{1/2}$ (n-InSb) and $\delta P_{02} \simeq 6 \cdot 10^{-12}$ (Ge:B) for the receiving elements. We note here that the receiving elements were not specially chosen for the measurements; obviously the figures presented for δP could be improved several times by choosing elements with the best threshold power.

In the final variant of the construction we made use of the fact that each of the two receiving elements is transparent in the spectral region of sensitivity of the other, and located them in a "sandwich" in a common Teflon cup with a wall thickness of ~0.5 mm (Fig. 1). Such fastening of the elements is more simple and convenient than fastening them independently to the walls of the chamber. Special measurements showed that in this variant the threshold powers are in no way worsened in comparison with the independent placement of the receiving elements.

LITERATURE CITED

1. A. E. Salomonovich and A. S. Khaikin, Tr. FIAN, 77, 33 (1974).
2. A. S. Khaikin, Tr. FIAN, 77, 56 (1974).
3. E. H. Putly, Proc. Phys. Soc., 73, 280 (1959).
4. T. M. Lifshits, Sh. M. Kogan, A. N. Vystavkin, and G. G. Mel'nik, Zh. Éksperim. i Teor. Fiz., 42, 959 (1962).
5. V. S. Ivleva, A. A. Kobzev, V. I. Lapshin, A. E. Salomonovich, V. I. Selyanina, V. F. Troitskii, A. B. Fradkov, and A. S. Kahikin, Preprint FIAN, No. 12 (1971).
6. V. M. Afinogenov, S. A. Aitkhozhin, V. A. Strakhov, A. A. Talegin, and V. I. Trifonov, Izv. Radiofiz., 15, 1572 (1971).
7. T. M. Lifshits, Vestn. Akad. Nauk SSSR, 12, 63 (1969).
8. T. M. Lifshits and F. Ya. Nad', Dokl. Akad. Nauk SSSR, 162, 801 (1965).
9. A. B. Fradkov and V. F. Troitskii, Tr. FIAN, 77, 85 (1974).
10. R. C. Ohlman, P. L. Richards, and M. Tinkham, J. Opt. Soc. Amer., 48, 531 (1958); W Witte, Infrared Phys., 5, 179 (1961); W. L. Eisenman, R. L. Bates, and J. D. Merriam, J. Opt. Soc. Amer., 53, 729 (1963); R. L. Bates and W. L. Eisenman, J. Opt. Soc. Amer., 54, 1280 (1964).
11. A. A. Sokolov, Optical Properties of Metals [in Russian], Fizmatgiz, Moscow (1961).
12. D. W. Williamson, J. Opt. Soc. Amer., 42, 712 (1952).
13. G. A. Morton and M. L. Schulz, RCA Rev., 20, 599 (1959).
14. A. N. Vystavkin, in: Semiconducting Devices and Their Application [in Russian], Sovetskoe Radio, Moscow (1968), No. 20.
15. K. Shivanandan, J. R. Houck, and M. Harwit, Phys. Rev. Lett., 21, 1460 (1968).
16. J. A. Jacquez, and H. F. Kuppenheim, J. Opt. Soc. Amer., 45, 460 (1955).

A CRYOGENIC SYSTEM CONTAINING LIQUID HELIUM FOR ON-BOARD RADIATION RECEIVERS

A. B. Fradkov and V. F. Troitskii

The creation of highly sensitive, low-inertia radiation receivers opens up the broadest prospects for the solution of a whole series of fundamental scientific and technological problems. The use of such receivers in extra-atmospheric studies on board space stations, artificial earth satellites, rockets, etc. has fundamental importance.

A powerful means of increasing the sensitivity of modern radiation receivers is their cooling to very low (1.8-5°K) so-called helium temperatures. Therefore the solution of the problem of cooling on-board radiation receivers to helium temperatures represents an urgent problem of great scientific and practical importance.

METHODS OF COOLING TO HELIUM TEMPERATURES
UNDER CONDITIONS OF ORBITAL FLIGHT

The cooling of objects (on board a spacecraft) to helium temperatures (1.8-5°K) can be accomplished at present by two basic methods:

a) the use of liquid helium obtained on the earth and retained on board the spacecraft in a special cryostat;

b) the use of an on-board refrigeration machine which produces the cold at the desired temperature level.

The use of liquid helium permits the simple and reliable temperature regulation of the cooled unit without requiring an additional energy supply on board and without introducing additional mechanical interference (oscillations, vibrations, etc.). A drawback of this method is the continuous consumption of the coolant to compensate for the heat inflows from the surrounding medium independent of whether the cooled unit is operating. The total time of temperature regulation is limited by the supply of liquid helium taken on board.

A refrigeration machine can be placed in operation as needed and its resource, determined by the wear of the mechanical units, can be very great. However, modern refrigeration machines for obtaining a temperature of 4.2-4.5°K have a number of serious drawbacks from the point of view of their use on board a spacecraft:

1) large energy requirement: about 1.5-3 kW or more of electrical power is required per watt of useful cold production [1];

2) large weight (~ 125-400 kg per watt of useful cold production [1]);

91

3) relative complexity of construction and preparation;

4) the necessity of complete dynamic balancing of the machine to prevent the appearance of undesirable oscillations and vibrations from the operation of the moving parts.

And while the development of on-board refrigeration machines at 4°K with characteristics better than those indicated above is now under way, satisfactory specimens of the machines have not yet been produced.

Therefore the method of using liquid helium for cooling the radiation receivers of an on-board telescope to 4-5°K was chosen as the more realistic for accomplishment in the near future.

The basis for the practical realization of this method is the creation of an on-board helium cryostat in which the liquid helium put in on the earth would be retained under flight conditions for the required time.

The conditions of space flight (weightlessness, g-forces, change in orientation, etc.) greatly complicate the problem of retaining and using liquid helium on board a spacecraft and demand the development of new methods and constructions.

Because of the small heat of vaporization (20 J/g) and the low boiling temperature (4.2°K at atmospheric pressure) the prolonged retention of liquid helium in general presents a difficult problem. The retention of small amounts (10-100 liters) of liquid helium on earth takes place in vessels with high-vacuum insulation and with shields cooled by liquid nitrogen. Such vessels lose to evaporation about 1-2% of the nominal volume of the liquid per day. Vessels with vacuum-multilayer insulation and with shields cooled by the evaporating gas are created for volumes of 50 liters or more. These vessels do not require additional nitrogen cooling but have somewhat higher losses to evaporation.

Modern helium vessels and cryostats work satisfactorily only under the condition that the exit neck joining the helium volume (T = 4.2°K) with the outer cover (T = 300°K) is placed vertically and the evaporating gas travels along it, removing most of the heat supplied through the heat conduction of the neck.

The neck must have a length of not less than 200-300 mm and have the minimum allowable diameter, or else the evaporation rate of the cryostat becomes extremely large. Since the volume of the gas at room temperature exceeds the volume of the liquid from which it is obtained more than 100-fold, the evaporating gas must be removed continuously to prevent the rupture of the cryostat.

In the operation of helium cryostats on board a spacecraft a number of additional difficulties arise in comparison with their operation on earth. The first difficulty is connected with the behavior of the liquid under the conditions of weightlessness. When cryostats with the ordinary two-phase state of the coolant (liquid–vapor) are used in space there is the danger that under the conditions of weightlessness the liquid will enter the exit neck, which leads to intense evaporation and the rapid emptying of the cryostat. This process is especially dangerous for liquid helium which has a very low heat of vaporization.

In the absence of the force of gravity the distribution of liquid in the vessel depends mainly on the ratio between the surface tension at the boundary of phase separation (liquid–vapor) and the surface forces between the liquid and the vessel walls.

The forces of adhesion to the walls are usually dominant for liquid coolants. This results in the liquid coating the entire inner surface of the cryostat while the gaseous phase lies within the liquid. In addition to the surface forces the shape of the vessel and the surfaces imbedded in it also have an important effect on the distribution of the coolant in the vessel under conditions of weightlessness.

For liquid helium the surface tension is many times smaller than for other coolants and therefore the use of various inserts can be quite effective.

Tests with liquid oxygen and hydrogen under conditions of weightlessness showed [2] that for the removal of gas from the vessel without loss of liquid, special measures and a device for the separation of the two phases are needed.

A second difficulty is connected with the presence of g-forces in the course of the active section of the flight and during movements of the craft in orbit. The occurrence of accelerations directed along the exit neck of the cryostat can result in the throwing of liquid into the neck and its rapid evaporation.

It must also be considered that in on-board cryostats the evaporating gas must be dumped off board into space containing a high vacuum.

And finally, it is undesirable to use liquid nitrogen for an additional coolant in on-board cryostats, as is usually done under laboratory conditions on earth.

A SCHEME FOR ON-BOARD CRYOSTATS

On the basis of the material presented above and with allowance for the experience obtained during tests in 1968 of the first helium cryostats on a satellite of the Kosmos series [3], an on-board cryostat for radiation receivers must satisfy the following conditions:

1) the cryostat must have a device (a "phase separator") to prevent the entrance of the liquid coolant into the exit neck during the space flight;

2) the cryostat must be equipped with a valve which automatically maintains a given pressure over the liquid helium with dumping of the evaporated gas off the craft;

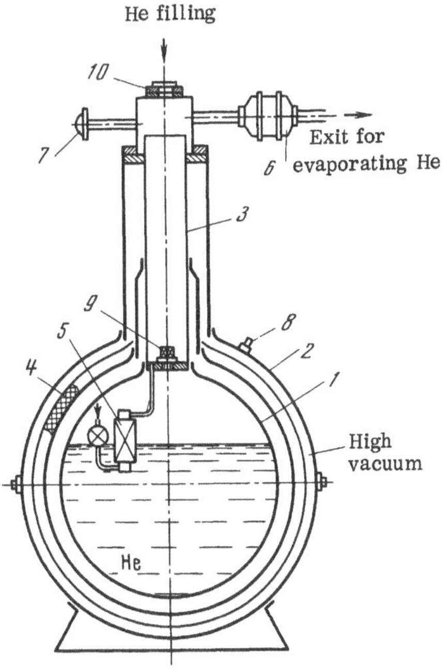

Fig. 1. Schematic diagram of on-board helium cryostat. 1) Volume for helium; 2) outer body; 3) tube-suspension; 4) shield-vacuum thermal insulation; 5) phase separator; 6) pressure regulator; 7) safety membrane; 8) valve for pumping out of thermal insulation; 9) plug for filling; 10) flange for insertion of object being cooled.

3) the cryostat must be rather simple and not require additional cooling by liquid nitrogen in flight.

A schematic diagram of an on-board helium cryostat is shown in Fig. 1. The helium volume 1 is suspended from the jacket 2 by the thin-walled suspension-neck 3 made of a low heat-conducting material (such as stainless steel). The helium volume is thermally insulated from the surrounding jacket with the help of a multilayered shield-vacuum insulator (SVTI) with an additional copper shield 4 cooled by the evaporating gas [4]. This makes it possible to avoid the use of liquid nitrogen. The preliminary filling with liquid helium takes place through an opening stopped by the plug 9.

After the plug is screwed in the helium volume 1 communicates with the neck 3 through the phase separator 5 which prevents the entry of liquid into the neck. Thanks to the phase separator only cold gas enters the neck in an amount corresponding to the liquid which has evaporated from the total inflow of heat to the helium volume. The evaporating gas in traveling along the neck cools the radiation receiver inserted in it as well as the shield 4 which is in thermal contact with the neck 3.

The pressure regulator 6 automatically maintains the assigned gas pressure in the neck (and consequently in the helium volume) independently of the pressure off board where the evaporated gas is dumped [5].

"LIQUID–VAPOR" PHASE SEPARATOR (PS)

Theoretically for the separation of the liquid and vapor phases one can propose several methods based on the differences in the physical properties of the liquid and vapor of the given coolant.

It follows from a preliminary analysis that the following systems of phase separation primarily deserve consideration:

a) mechanical systems based on the difference in densities of the vapor and liquid and using a rotating element which exerts centrifugal forces on the two-phase mass, thus separating it and passing only the vapor to the discharge;

b) dielectrophoretic methods based on the different polarization of the phases in a non-uniform electric field; if a two-phase mixture is passed, for example, through an electrostatic capacitor with a high voltage the liquid will be moved preferably to regions of high field strength;

c) systems using surface tension forces to control the boundaries of liquid–vapor separation;

d) heat-exchange-refrigerator systems in which the two-phase mixture being dumped from the reservoir is throttled to lower pressures and temperatures and passed through a heat exchanger to evaporate the liquid present in the mixture.

In [6] a comparative survey is made of the possible methods listed above for the separation of the liquid and vapor phases in application to a rocket fuel such as hydrogen. It follows from the survey that the thermal method of phase separation is the most preferable. We also came to the same conclusion in an examination of the different methods of separating the vapor and liquid phases for helium.

Several types of heat-exchange-evaporator phase separators for cryostats containing liquid helium were developed and studied. A description of one of the separators developed is given below.

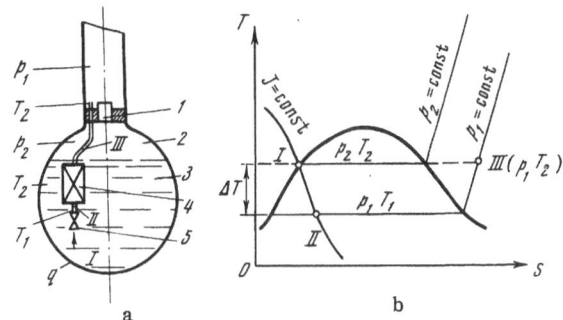

Fig. 2. Diagram of connection of phase separator (a) and
representation of the process in a T−S diagram (b).   1)
Plug; 2) vapor; 3) coolant; 4) heat exchanger; 5) throttle.

The separator consists of a heat exchanger of copper tubing coiled into a spiral with a
throttle device at the input end.

The heat exchanger with the throttle device is placed within the helium volume of the
cryostat and its other end is connected to the exit neck of the cryostat (Fig. 2).

The phase separator operates as follows: after the cryostat is filled with liquid helium
and stationary conditions are established (with a pressure p_1 within the cryostat and in the
exit neck and a coolant temperature T_1) the opening to the helium volume through which the
filling took place is blocked with the plug. Now the outflow of coolant from the vessel into the
exit neck is possible only through the phase separator. After the plug is inserted the pressure
within the volume is raised to a certain value p_2 and the temperature accordingly to T_2 because of the presence of heat flow into the volume containing the coolant. Under the effect of
the pressure difference $(p_2 - p_1)$ the coolant overcomes the resistance of the throttle of the
phase separator and enters into the tubing of the heat exchanger. Here as a result of the
throttling process (see the T−S diagram in Fig. 2) the pressure of the coolant which entered
the separator falls from p_2 to near p_1 and the temperature falls from T_2 to T_1. Because of the
temperature difference $\Delta T = T_2 - T_1$ formed the liquid phase within the separator evaporates
and only vapor enters the exit neck. The surface of the heat exchanger must be such that
evaporation of the entering liquid is assured at the small difference (on the order of 0.1-0.2°)
between the temperatures T_2 and T_1.

During the operation of the separator the following situations can arise in practice:

1) liquid coolant enters into the separator, while on the outside it is also washed by
liquid;

2) liquid coolant enters the separator, while outside it is surrounded by saturated vapor;

3) saturated vapor enters the separator, while outside it is surrounded by liquid coolant;

4) saturated vapor enters the separator, while outside it is surrounded by saturated
vapor.

Situations 3 and 4 are not subject to consideration since in this case saturated vapor will
enter the suspension-neck, which is what we require. Situation 2, in which the total heat transfer coefficient will be less than in situation 1, is the most important from the aspect of heat
exchange.

The situations considered can be obtained and experimentally tested under stationary conditions. Under conditions of alternating accelerations all the situations enumerated are to a certain extent transient, but situation 2 can play the main role. With weightlessess, in the case of situation 2 the heat exchange from the side of the saturated vapor is accomplished by thermal conduction instead of natural convection, which reduces the total heat transfer coefficient and requires an increase in the heat exchange surface. However, a design solution can be found allowing one to avoid situation 2 and improve situation 1.

It is seen from Fig. 1 that the heat flux q from the surrounding medium arrives first at the surface of the volume for the coolant and only then is transferred to the coolant. By soldering the heat exchanger of the separator to the inner wall of the coolant volume we obtain a "solid-boiling liquid" heat-exchange system, which is considerably more favorable than the "liquid (or saturated vapor) —wall − liquid" system. It should be noted that the wall temperature of the helium volume is higher than the coolant temperature, if only slightly.

Thus, for greater definiteness of the conditions of heat exchange in the different situations and to increase the reliability of operation the heat exchanger of the phase separator or part of it should be soldered to the inner surface of the coolant volume.

The required surface S of the heat exchanger is estimated from the equation

$$S = \frac{q}{\alpha \Delta T},$$

where q is the heat flux to the coolant volume (determined from the thermal design of the cryostat), α is the coefficient of heat release from the wall to the boiling coolant, and ΔT is the temperature difference between the wall and the coolant (taken in the range from 0.1 to 0.2°).

The area F of the through cross section of the throttle at the entrance to the heat exchanger of the separator is determined from a well-known equation of hydraulics

$$F = \frac{G}{\psi W \gamma_l},$$

where G is the weight flow rate of the liquid (the evaporation rate of the cryostat) in kg/sec, $G = q/r$ (r is the heat of vaporization of the coolant); ψ is the flow rate coefficient, γ_l is the density of the liquid; W is the discharge rate of the liquid in the throttle opening:

$$W = \sqrt{2g \frac{p_2 - p_1}{\gamma_l}}.$$

In good helium cryostats the rate of helium evaporation is about 4-8 g/h, and for such flow rates the calculated diameter of the opening in the throttle is obtained as several dozen microns. It is extremely difficult to create an automatically regulating throttle with small openings.

Various devices can be used as nonregulating throttle elements in our case: a throttle disk with a specially machined opening, a capillary, a valve-seat system, narrowing of a tube, etc. After tests of various types of throttles the narrowing of a tube proved to be the most suitable method of obtaining the required size for the throttle gap.

The preparation of a throttle by narrowing a tube is accomplished on a press. It is necessary to note that the narrowing of the tube can be done at several places (a multistage throttle), thereby obtaining gaps of larger size than a single gap and thus decreasing the possibility of clogging the throttle. The possibility of attaining a resistance of the desired value is also an advantage of a multistage throttle. After fabrication the throttle is calibrated in two

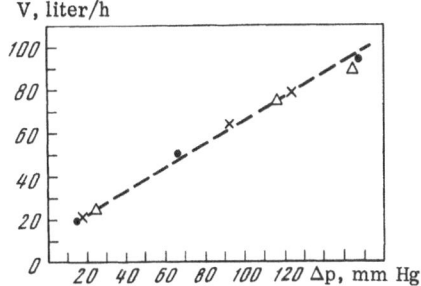

Fig. 3. Dependence of flow rate of gas through phase separator on pressure drop.

stages. First calibration is carried out at room temperature on gaseous helium. The calculated flow rate of the helium escaping through the throttle is checked experimentally. After the calculated and actual flow rates become similar (through a change in the gap area) one goes on to the second stage of calibration on liquid helium. For this the throttle is connected to the heat exchanger and placed in a cryostat where the actual working conditions of the phase separator are imitated.

By varying the pressure above the liquid helium entering the separator and measuring the amount of gas emerging from the separator a graph is taken of the dependence of the flow rate [liters/h (STP)] through the separator on the pressure drop on it.

The quality of the phase separator, i.e., the absence of traces of liquid at the exit, is tested by two methods: visually through a window with bias lighting and using a resistance thermometer sensitive to the difference between the liquid and vapor phases. A typical characteristic curve for a separator is presented in Fig. 3.

CONSTRUCTION OF ON-BOARD HELIUM CRYOSTATS

Besides the presence of a phase separator, an on-board cryostat differs in the design of the exit neck. An ordinary neck, consisting of a thin-walled tube 200-300 mm long with an open cross section, is poorly suited for an on-board cryostat since convection currents will develop in the gas filling such a neck during changes in the position of the cryostat in space. These convection currents are determined by the difference in gas densities at the ends of the neck and depend on the acceleration of gravity.

When the cryostat is turned 180° (neck down) under laboratory conditions on earth the helium evaporation rate increases because of convection severalfold in comparison with the neck up position of the cryostat. A special neck construction was developed for on-board cryostats to decrease the effect of convection on the evaporation rate (Fig. 4).

The neck consists of two coaxial tubes between which a coil of stainless wire is compactly wound. After the separator the gas enters the space between the tubes and moves along the spiral toward the warm exit end. Such a device makes it possible to sharply decrease the cross section and to lengthen the path for the outflow of the gas. The inflow of heat to the helium volume due to convection when the force of gravity is directed from the helium volume to the neck is reduced as a result.

The use of a double-walled neck with a spiral insert increases the durability and rigidity of the suspension of the helium volume from the jacket, which is an additional advantage. A structural diagram of one of the types of on-board cryostats is presented in Fig. 4. The cylindrical helium volume 1 with the phase separator 3 inside it is suspended from the outer jaceket

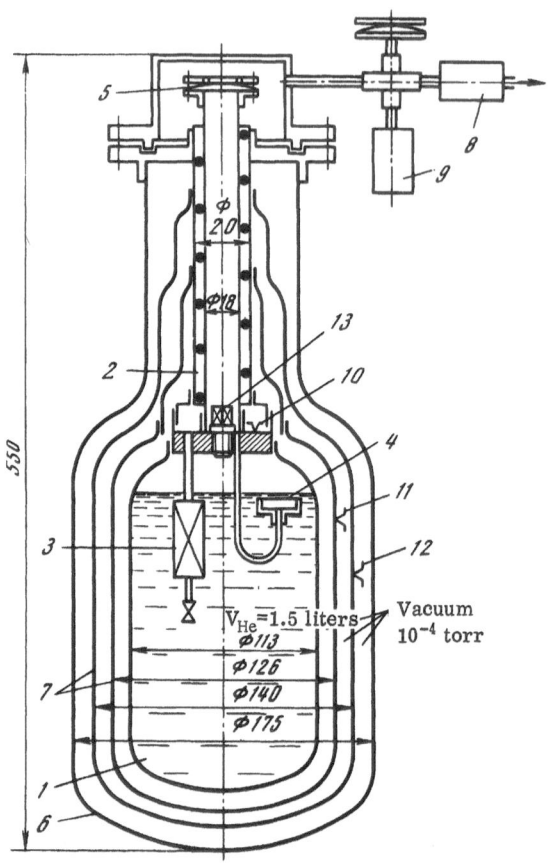

Fig. 4. Structural diagram of cryostat with double-walled neck. 1) Helium volume; 2) neck; 3) phase separator; 4, 5) safety membranes; 6) body of cryostat; 7) shields; 8) pressure regulator; 9) pressure pickup; 10-12) resistance thermometers; 13) plug.

through the double-walled neck 2 to which two copper shields 7 are fastened. The shields are cooled by the evaporating gas passing in the spiral gap between the tubes of the neck. The high vacuum and the wrapping of the shields with an aluminum-coated Dacron film provide sufficiently effective thermal insulation.

The system of safety membranes 4, 5 protects the cryostat from rupture in the case of failure of the phase separator. The radiation receiver being cooled can be mounted in the inner tube of the neck or inserted through the plug 13 into the helium volume. The pressure within the helium volume (and consequently the temperature) is established and maintained by the pressure regulator 8 through which the dumping off board of the evaporated gas takes place.

LABORATORY AND SPACEFLIGHT TESTS

OF MODELS OF ON-BOARD CRYOSTATS

The cryostat models fabricated are subjected to tests for durability, tightness, and evaporation rate under laboratory conditions before mounting on board. The liquid helium evaporation rate and the operation of the phase separator are tested in two positions — normal (neck up) and turned 180° (neck down).

The results of laboratory tests of cryostat KKR-4, built according to the structural diagram of Fig. 4, are presented in Fig. 5. The geometrical volume of the cryostat allows it to be filled with about 1.5-1.6 liters of liquid helium. The designed evaporation rate is ~2.6 g/h, which provides the maintenance of a helium temperature for about three days.

As the tests showed (Fig. 5), the time for the total evaporation of the helium is ~80 h for the normal cryostat position and ~62 h in the neck-down position. In the first case the shield temperatures were 133 and 223°K and in the second case 96 and 200°K, respectively.

In 1971 the cryostat KKR-4 was tested on an earth satellite of the Kosmos series. According to the telemetry data obtained, presented in Fig. 6, there was still liquid helium in the cryostat on the 36th revolution of the satellite (56 h after the launch). There was no more helium on the 46th revolution. Thus, the total evaporation of helium occurred between the 36th and 46th revolutions. By comparing the nature of the warming of the KKR-4 shields in orbit

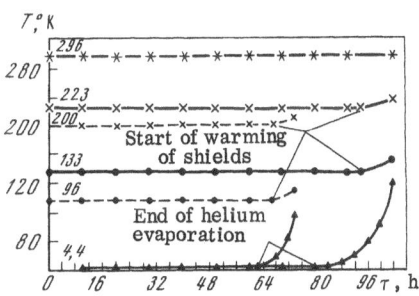

Fig. 5. Working temperatures of cryostat KKR-4 under laboratory conditions. Solid lines: neck turned upward; dashed lines: vertically downward.

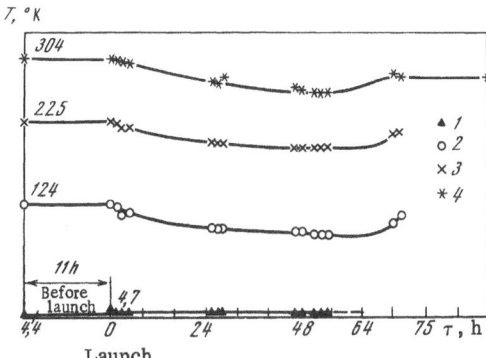

Fig. 6. Temperatures in cryostat KKR-4 during orbital flight. 1) Liquid helium; 2) shield I; 3) shield II; 4) outer jacket.

and in the laboratory and allowing for the time from the filling to the launch ($\sim$11 h) one can compute that the total time of helium retention in the cryostat is more than 72 h, including more than 60 h under the conditions of orbital flight.

From the information obtained it is seen that the helium temperature in flight was 4.7°K, higher by 0.3° than under prelaunch conditions. The increase in the boiling temperature of the helium is explained by the turning on of the pressure regulator which maintained in flight a pressure somewhat exceeding atmospheric pressure.

The laboratory and spaceflight testing of cryostat KKR-4 confirmed the correctness of the principles and constructional solutions of the units of an on-board cryostat for the retention of liquid helium under the conditions of orbital flight.

LITERATURE CITED

1. Technology of Liquid Helium, NBS Monograph No. 111, New York (1968).
2. New Directions in Cryogenic Technology [Russian translation], Mir, Moscow (1966).
3. A. B. Fradkov, V. F. Troitskii, G. N. Mikhailova, N. N. Inozemtsev, and V. V. Svistel'nik Pribory i Tekh. Éksperim., No. 6, 205 (1969).
4. A. B. Fradkov, Dokl. Akad. Nauk SSSR, 133, 82 (1960).
5. N. I. Inozemtsev and G. N. Mikhailova, Zavod. Lab., 36, 752 (1970).
6. R. C. Mitchell, I. A. Stark, and R. C. White, Adv. Cryogen. Eng., 12, 72 (1967).

BAND-PASS FILTERS FOR THE SUBMILLIMETER RANGE

S. V. Solomonov, O. M. Stroganova, and A. S. Khaikin

INTRODUCTION

In the solution of problems of submillimeter astronomy it often becomes necessary to record signals from a broad spectrum using a nonselective radiometric apparatus whose receivers are also broad-band. At the same time it is always desirable to obtain information on the spectral composition of the radiation received. Obviously the problem of spectral analysis can be solved in this case by placing in front of the receiver changeable, relatively narrow-band elements (band-pass filters) which transmit radiation in different sections of the selected range.

These filters must have small losses in the transmission band and should transmit practically no radiation outside the limits of this band. The width of the transmission band of the filters is determined by the specific problem. A natural limitation on the resolving power (the quality $Q = f/\Delta f$) of the filters is the condition of the reliable recording of the signal against the noise background. The latter in turn depends on the source of the radiation and on the basic parameters of the radiometer (fluctuation sensitivity, directional diagram of antenna, etc.).

One can mention problems where filters with relatively low Q are sufficient. For example, the relict background radiation is distinguished by the extremely low level of the power received. Since one of the important problems in its recording is the determination of the nature of the radiation, characterized by the overall form of the spectrum, high resolution is not required and it proves possible to conserve the power of the radiation supplied to the receiver by choosing filters with low quality. Filters with relatively low Q are sometimes also necessary for the efficient operation of spectral instruments with high resolution: Under these cases the filters perform the preliminary monochromatization (to cut out harmonics, etc.).

Thus, band-pass filters with qualities of from a few units to several tens or more are of great interest from the aspect of the development of spectroradiometer.

1. METHODS OF RADIATION FILTERING AND PROPERTIES OF THE OPERATION OF FILTERS UNDER ON-BOARD CONDITIONS

The following basic types of filters for the submillimeter range which have a marked dispersion of the transmission or reflection coefficient are known at present: absorption filters based on optical materials, power filters, filters based on periodic structures.

The optical materials used as filters in the submillimeter range have characteristic curves which possess relatively flat fronts, which in practical systems of monochromators

101

leads to the use of filters of these materials only in conjunction with other filtering elements [1, 2].

Artificial materials — composites based on powders of alkali-halide salts and the oxides of certain metals [3] — have characteristic curves of transmission with considerably steeper fronts than for ordinary optical materials. However, such filters are sufficiently efficient only in the short-wave region of the submillimeter range (to about $\lambda \leq 200 \, \mu$). This is connected with the absence of suitable materials providing absorption with a steep drop in the longer wavelength region.

Periodic structures — echelette gratings [1, 4], grids and structures complementary to them [5] which work both as reflecting and as transmitting filters, interference filters [6], and multielement structures [7] — permit the rather effective solution of the problem of radiation filtering in almost the entire submillimeter range.

In the development of on-board spectroradiometers, other conditions being equal, preference must obviously be given to filtering elements which do not significantly increase the overall size and weight of the instrument (which can be placed, for example, in the converging beam directly behind the objective), are able to withstand mechanical loads, and can be used under conditions of an external vacuum and sharp temperature drops without a marked change in spectral characteristics. In addition, rigid sensitivity requirements are imposed on the radiometric apparatus, as a rule, which leads to the necessity of using input elements having minimum losses. In the majority of cases, when unpolarized radiation is recorded, the radiometer must possess as far as possible the same spectral sensitivity with respect to any polarization of the radiation received. These rigid requirements, as follows from a comparison of the types of filters enumerated, are satisfied to the greatest extent by the third type. The results of studies of the spectral characteristics of metallic screens having periodic perforation, which permit the most convenient construction of band-pass filters for the submillimeter range, will be presented below.

2. TRANSMISSION OF METALLIC SCREENS HAVING

PERIODIC PERFORATION

Let us examine the properties of metallic screens of finite thickness h having periodic perforation and operating on transmission in the region of wavelengths λ on the order of the screen period g (Fig. 1). Such screens with $2a \ll g$ and $h \ll \lambda$ are usually [5, 7-9] called induction grids. The transmission of a plane electromagnetic wave through an infinitely thin metallic screen having periodic perforation for different forms of openings and through a screen with a dielectric coating is examined theoretically and experimentally in [5-16]. Measurements [8] and calculations [9] of the transmission of thin metallic induction grids show that an increase in the transmission T occurs in the region $\lambda \approx g$ while the transmission decreases outside this region. For a qualitative explanation of the behavior of the transmission curve Vogel and Genzel [8] consider a metallic two-dimensional grid in the region of $\lambda < g$ in the form of two identical crossed diffraction gratings with a spacing $g' = g$ placed one behind the other in the path of a pencil of rays. In this case the directions toward the diffraction maxima, determined by the angles β_m, can be described by the well-known equation

$$\lambda = g' \frac{\sin \alpha - \sin \beta_m}{m}, \qquad m = 0, \pm 1, \pm 2 \ldots,$$

where g' is the lattice spacing, α is the angle of incidence, and β_m is the angle corresponding to the direction toward the diffraction maximum of m-th order. It is obvious that the condition of the disappearance of diffraction at $\alpha = 0$ is $m = 1$, $\beta_m = \pm 90°$. In this case the radiation in the first diffraction maximum slips along the surface of the diffraction lattice. Thus, the boundary of the disappearance of diffraction corresponds to the ratio $\lambda / g = 1$. With an increase in the angle of incidence α, as follows from the equation, the boundary of the diffraction losses shifts into the long-wave region. The appearance of strong reflection should be expected in the region of $\lambda > g$. Actually, the extreme long-wave point of this region is $\lambda = \infty$ (direct current) where the grid represents a perfect conductor, i.e., a perfect mirror. With a decrease in λ the reflection should decrease until in the region of very short waves it starts to obey the laws of geometrical optics (the reflected power is proportional to the ratio of the total area of the opaque sections to the total area of the openings).

This qualitative analysis does not, however, provide an explanation for certain effects which develop when a plane electromagnetic wave falls on a grid with $\lambda \approx g$: first, the fact of the displacement of the transmission maximum in the long-wave direction remains unexplained; second, the processes leading to the fact that the almost total absence of reflected and diffracted radiation power is observed at the point of maximum transmission are unclear.

Theoretical methods developed in recent years for the calculation of diffraction scattering on periodic structures permit one to calculate the dependence of the transmission coefficient on the wavelength and the parameters of perforated screens. A perfectly conducting infinitely thin screen is examined in [10-16] using the Ritz–Galerkin variational methods. The problem of diffraction on a thin perforated screen is reduced to the case of the propagation of a wave in an equivalent line.

A thick screen with dielectric filling of the openings of rectangular shape was examined in [17]. In this case instead of the perforated screen its individual "cell," which is an element of the periodic structure, is considered. The cell can be considered as a "thick" diaphragm in some fictitious waveguide. Starting from the equations of continuity of the tangential components of the magnetic field strength at the boundaries of the region and adopting the well-known technique of the Galerkin variational method one can obtain an expression for the input admittance Y of the screen normalized to the wave impedance of free space:

$$Y = j(B_p + B_b) + \frac{B_{mu}^2}{1 + j(B_p + B_b)} = G + jB,$$

where B_p is the susceptance for higher types of waves from one side of the screen, B_b is the intrinsic susceptance, and B_{mu} is the mutual susceptance.

Then for the modulus of the transmission coefficient we obtain

$$|t| = 2\sqrt{\frac{G}{(1 + G)^2 + B^2}}.$$

In comparing the results of calculations of the transmission coefficients of different perforated screens ("thick" and "thin") without allowance for ohmic losses in them one can draw the following general conclusions.

1. The incidence of a plane electromagnetic wave of the TEM type on a screen leads to the excitation at its surface of higher types of waves, which die out with departure from the screen, forming around it a certain space where part of the incident energy is stored in the form of electric and magnetic fields. For the description of the transmission and reflection of the screen one can introduce the concept of the equivalent input admittance, characterized by the fraction of energy stored in higher types of oscillations. The passage of radiation through a screen having periodic perforation has a resonance nature.

2. The high transmission when $\lambda \approx g$ is caused by resonance on the screen period, which must be distinguished from possible resonances on the apertures.

3. An increase in the thickness of a screen having rectangular openings with dielectric filling leads to narrowing of the resonance curves of transmission [17] without changing its other parameters.

3. STUDY OF TRANSMISSION OF SCREENS HAVING

SQUARE OPENINGS WITHOUT DIELECTRICS

The calculations and measurements of the transmission of screens when $\lambda \approx g$ which have been conducted up to now by various authors pertain mainly to thin perforated screens. The transmission of "thick" screens [17] is calculated only for the case of a screen having rectangular openings with dielectric filling. We have studied experimentally the transmission of "thick" screens having periodic perforation without special filling. The screen openings had the form of a square (Fig. 1) with $\varepsilon = 1$. The basic parameters of the screens studied are presented in Table 1. Measurements of the transmission of the screens were conducted in a broad range of wavelengths, from 50 to 1000 μ, on a Hitachi FIS-21 spectrometer with normal incidence of the radiation and with angles of beam convergence of about 10°.

Screens produced by the domestic industry and modifications of these screens obtained from the industrial screens by polishing them or by galvanic accretion of copper or nickel to the desired values of the parameters $p = h/g$ and $p = 2a/g$ (see Fig. 1) are presented in Table 1. These modifications are marked with an asterisk in the table. The power transmission coefficients as functions of the wavelength for these screens are presented in Figs. 2 and 3. The measurements showed that the transmission of the screens has a resonance nature with the transmission maxima T_{max}, comprising 62-90% at the wavelengths λ_{max} located in the regions $\lambda_{max} = (1.05-1.16)g$. Relatively narrow "dips" appeared on the $T(\lambda)$ curves of some of the screens, evidently indicating the presence of wave guide types of oscillations in the "thick" screen and the possibility of their interference as a result of reflections from the front and back planes of the screen. This fine structure is not shown on the transmission curves pre-

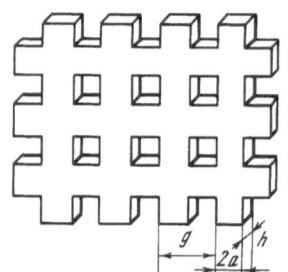

Fig. 1. Configuration of perforated screen.

TABLE 1

Screen No.	g, μ	$2a$, μ	h, μ	Q_l	Screen No.	g, μ	$2a$, μ	h, μ	Q_l	Screen No.	g, μ	$2a$, μ	h, μ	Q_l
8	230	95	100	4.1	103 *	500	220	250	3.6	402	100	25	35	2.1
9	230	100	120	5.0	104 *	500	220	225	3.1	403	100	25	30	1.8
3	230	105	135	5.1	105 *	800	220	360	1.3	405	165	75	90	5.4
25	500	120	120	1.1	106 *	500	210	190	4.0	406	180	70	90	5.3
100 *	500	210	230	3.4	400	130	60	80	4.7	408 *	165	75	77	4.2
102 *	500	170	180	2.8	401	100	22	32	1.8	409 *	130	60	60	5.3

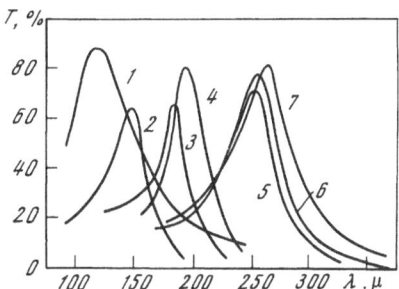

Fig. 2. Transmission of industrial screens.
1) Screen 401; 2) 400; 3) 405; 4) 406; 5) 3;
6) 9; 7) 8.

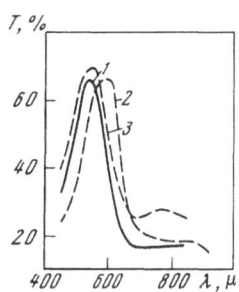

Fig. 3. Transmission of modifications of industrial screens. 1) Screen 103; 2) 100; 3) 102.

sented (Figs. 2 and 3) since the effect of the "dips" on the resultant characteristics of the filters is insignificant. As seen from Table 1, there is a large scatter in the values of the quality Q_l (from 1.1 to 5.4), which is connected with the differences in the geometrical parameters of the screens. For the comparison of screens with different thicknesses, periods, and resonance wavelengths let us introduce the values λ/g, $q = 2a/g$ (the filling coefficient), and $p = h/g$ (the relative thickness). Graphs of the dependence of the quality of the screens on their relative thickness are presented in Fig. 4. These graphs show that at a fixed q the quality of the screen increases with an increase in thickness. The graphs presented in Fig. 4 pertain to screens with p = 0.25 and 0.44. For $Q_l \leq 1$ the experimental points were borrowed from [7, 18, 19]. The U-R-G point from [18] is incorporated directly in the Q_l (p) curve for q = 0.25. The R-M point from [7] and the T point from [19] show that the $Q_l(p)$ curve for q = 0.44 issues from the region close to the origin of coordinates. The calculated $Q_l(p)$ functions obtained by Panchenko [17] for screens with rectangular openings filled with dielectric with $\varepsilon = 4$ and $\varepsilon = 2.5$ are plotted on the same graph. As is seen, the theoretical curves have a lesser slope for the smaller ε. A sharp increase in the quality occurs with an increase in the screen thickness. Thus, one can conclude that there is qualitative agreement between the $Q_l(p)$ curves which we obtained for square openings and the curves calculated in [17]. Experimental Q_l (p) functions are presented in Fig. 5. The lower curve corresponds to the "thin" screens (grids) known from [5, 7, 8, 18]. All these screens are characterized by small values of p (compared with the relatively thick screens 104, 105, and 405) and therefore the experimental points corresponding to "thin" screens were arbitrarily joined by a single curve. As seen from the figure, a sharp increase in the quality with an increase in q occurs only for "thick" screens.

Thus, the study of perforated screens shows that their transmission characteristics are determined by the geometrical parameters p and q. The experimental dependences of the

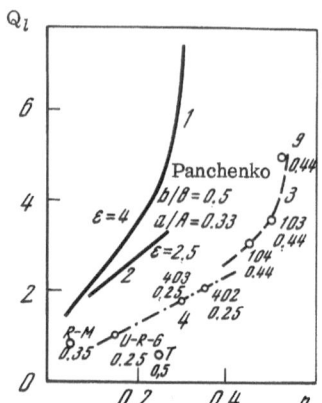

Fig. 4. Dependence of quality of different
screens on their relative thickness.

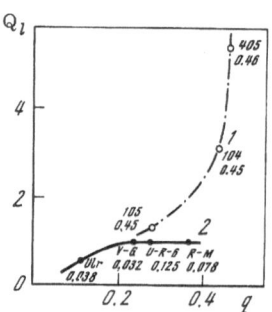

Fig. 5. Experimental loaded qualities for thick
(1) and thin (2) screens [5, 7, 8, 18].

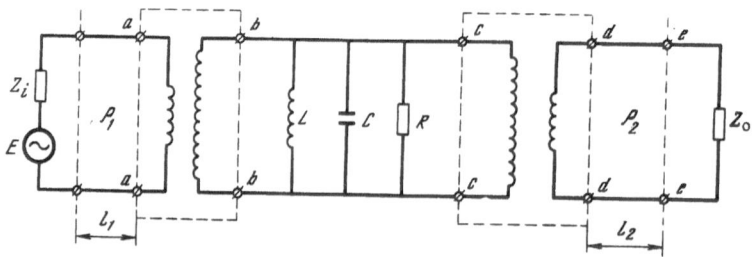

Fig. 6. Equivalent line in the problem of the transmission of thick perforated
screens.

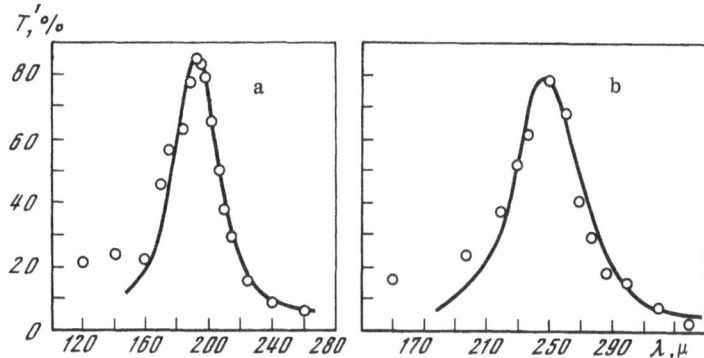

Fig. 7. Comparison of experimental characteristics of thick screens with those calculated by the method of equivalent lines. a) Screen 9; b) screen 406. Points: experiment; lines: calculation.

quality Q_l on these parameters obtained make it possible to develop band-pass filters for the submillimeter range with the assigned transmission characteristics on the basis of such periodic structures.

An equivalent circuit for a thin screen in the form of a parallel oscillatory circuit containing losses is proposed in [5] to describe the operation of thin induction grids. In the case of "thick" screens, as calculations have shown, the transmission in the region of $\lambda \gtrsim g$ can be described rather well using the equivalent circuit presented in Fig. 6. Here the "thick" screen is given in the form of a resonance circuit containing losses connected with the input and output lines using ideal transformers. In Fig. 7 the experimental data for the transmission of screens 9 (a) and 406 (b) are shown by points and the corresponding calculated transmissions for the equivalent circuits of these screens are shown by solid lines.

The use of these equivalent circuits is not legitimate in the region of $\lambda < g$. An analysis of the equivalent circuits of "thick" screens shows that a change in the filling coefficient q of the screen or in the relative thickness p corresponds to a change in the coupling of the equivalent resonator with the input and output lines. In this case an increase in q (p fixed) or an increase in p (q fixed) leads to a decrease in the coupling, which increases the loaded quality Q_l of the resonator and decreases T_{max}.

It follows from the well-known [20-23] equations of the theory of long lines that the quality Q_l, which can vary with a change in the coupling, is limited from above by the value of the intrinsic quality Q_0 of the resonator which, as estimates show, can be on the order of 10^2 for the "thick" screens listed in Table 1.

By using several such resonators connected in series with uncouplings between them one can obviously obtain a circuit with a resultant transmission equal to the product of the transmission coefficients of the separate resonators forming it. In this case one should expect a larger transmission coefficient at the resonance wavelength with considerably steeper slopes and stronger quenching outside the transmission band of the circuit. Below we will examine the attainment in practice of the series arrangement of resonators for the construction of a band-pass filter using "thick" metallic screens having perforation.

4. DEVELOPMENT OF FILTERS BASED

ON "THICK" METALLIC SCREENS

In the construction of filters composed of "thick" screens arranged in series it is necessary to eliminate interference effects which can lead to the appearance of "parasitic" peaks on
the transmission curve. For this the screens are set at a small angle (3-5°) to one another.
Measurements carried out on a Hitachi FIS-21 spectrophotometer with beam convergence angles of about 16° showed that the characteristic curves of the screens are multiplied under
these conditions. As an illustration of this the transmission curves of screens 401 and 8 are
presented in Fig. 8. Curves 1 correspond to the transmission of single screens while curve
2 in Fig. 8a corresponds to the resultant transmission of two 401 screens placed at a spacing
of about 4 mm from one another. The transmission of three 8 screens located one after
another at a spacing of about 5 mm is presented in Fig. 8b (curve 2).

A decrease in the spacing between these screens leads to some increase in the transmission in the short-wave region, which is evidently connected with the falling of part of the diaphragm power into the field of view of the receiver. With spacings between screens equal to
about 2 mm, however, the transmission in this region increases insignificantly with an unchanged value of T_{max}. A filter using identical 8 screens mounted one after another at a spacing of 2 mm was fabricated on the basis of these preliminary measurements. The characteristic curve of transmission of this filter measured in the region from 30 to 500 μ is presented
in Fig. 9 (curve 1). The basic parameters of the filter obtained are: λ_0 = 260 μ, T_{max} = 53%,
Q_l = 7.4, transmission outside band

$$T \leqslant \begin{cases} 1.5\% & \text{for } \lambda < 190 \ \mu, \\ 1\% & \text{for } \lambda > 300 \ \mu. \end{cases}$$

The filter obtained is simple in construction, does not require complicated tuning and
alignment, has characteristics which are critical toward moderate changes in the angle of incidence of the radiation or the spacings between the screens, operates at beam convergence
angles of up to 16°, and has a small overall size and weight. This makes it possible to mount
it in spectral radiometers in the converging beam right behind the objective. Since the screens
are made of nickel the filter can operate under conditions of high humidity. The properties of
the filter indicated above allow it to be used in a wide class of spectral radiometers: cooled

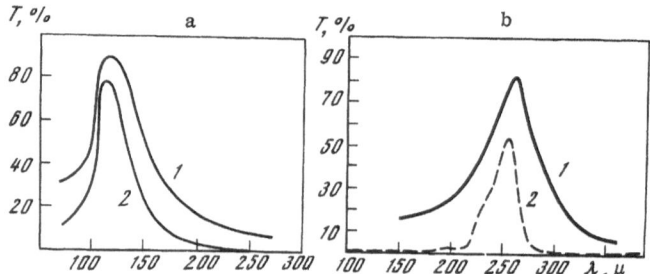

Fig. 8. Characteristic curves of transmission of single and of several
thick screens placed one after another. a) Screen 401: 1) single; 2)
two screens at a spacing of 4 mm; b) screen 8: 1) single; 2) three
screens at a spacing of 5 mm each.

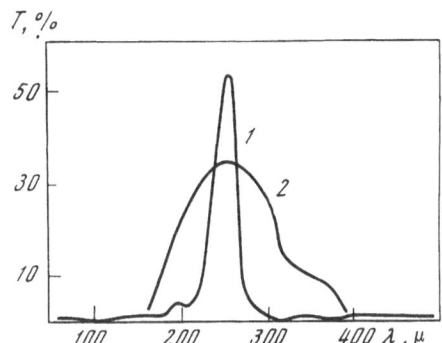

Fig. 9. Characteristic curve of a filter of three screens at a spacing of 2 mm from one another (1) compared with the combination filter of [21] (2).

and uncooled, laboratory and on-board — under conditions of temperature and pressure differentials and in the presence of mechanical loads.

For a comparison with the types of filters developed earlier in the same wavelength range the transmission of a filter described in [21] is presented in Fig. 9 (curve 2). This filter is made using a combination of four metallic grids and one Yamada power filter. It is seen from the figure that the transmission curve of this combination filter has T_{max} = 35% at λ_0 = 260 μ, relatively flat slopes, and a lower quality.

CONCLUSION

The results of the calculations and measurements presented above indicate the possibility in principle of the development of band-pass filters for the submillimeter range using the effect of the resonance transmission of perforated screens in any region of the range with a relative transmission band from several dozen percent to several percent.

A specimen of a band-pass filter of the submillimeter range suitable for use in the composition of an on-board submillimeter telescope was developed and tested as a result of the studies conducted. The required filters for the entire range of working wavelengths can be built on the basis of the type of filter developed. The concrete parameters of the filters must be determined with allowance for the program of studies and the expected sensitivity of the receivers.

The authors express their deep appreciation to A. E. Salomonovich for constant attention to the work and to R. I. Perets for a useful discussion.

LITERATURE CITED

1. N. G. Yaroslavskii, Doctoral dissertation, State Optics Institute, Leningrad (1965).
2. Techniques of Spectroscopy in the Far Infrared, Submillimeter, and Millimeter Regions of the Spectrum [Russian translation], D. Martin, ed., Mir, Moscow (1970), Chap. 3.
3. Y. Yamada, J. Opt. Soc. Amer., 52, 17 (1962).
4. É. A. Tishchenko, Opt. i Spektr., 30, 159 (1971).
5. R. Ulrich, Infrared Phys., 7, 37 (1967).
6. R. Ulrich, Infrared Phys., 7, 65 (1967).
7. G. M. Ressler and K. D. Möller, Appl. Opt., 6, 893 (1967).
8. P. Vogel and L. Genzel, Infrared Phys., 4, 257 (1964).
9. B. D. Saksena, D. R. Pahwa, M. M. Pradhan, and K. Lal, Infrared Phys., 9, 43 (1969).
10. R. B. Kieburtz and A. Ichimaru, IRE Trans., AP-9, 6 (1961).

11.   L. Oh, C. D. Lunden, and C. Chiou, Microwave J., 7, 62 (1964).
12.   C. C. Chen, IEEE Trans., MTT-18, 627 (1970).
13.   C. C. Chen, IEEE Trans., MTT-18, 475 (1971).
14.   B. A. Panchenko, Izv. Vuz. Radiofiz., 13, 465 (1970).
15.   B. A. Panchenko, Izv. Vuz. Radiofiz., 11, 1884 (1968).
16.   B. A. Panchenko and I. P. Solov'yaninov, Izv. Vuz. Radiofiz., 13, 467 (1970).
17.   B. A. Panchenko, Radiotekhn. i Élektron., 12, 719 (1970).
18.   R. Ulrich, K. F. Renk, and L. Genzel, IEEE Trans., MTT-11, 363 (1963).
19.   É. A. Tishchenko, Candidate's dissertation, Institute of Physical Problems, Moscow (1971).
20.   R. I. Perets, Some Problems of the Development of Antenna Switches [in Russian], Sovetskoe Radio, Moscow (1951).
21.   S. P. Varma and K. D. Möller, Appl. Opt., 8, 2151 (1969).
22.   I. V. Lebedev, UHF Techniques and Instruments [in Russian], Vol. 1, Moscow (1970).
23.   A. F. Harvey, UHF Techniques [Russian translation], Vol. 1, Sovetskoe Radio, Moscow (1965).

PROPERTIES OF THE CONSTRUCTION OF AN
ON-BOARD SUBMILLIMETER TELESCOPE

V. N. Bakun, P. D. Kalachev, A. E. Salomonovich,
and A. S. Khaikin

The specific properties of the construction of telescopes, including submillimeter telescopes, intended for extra-atmospheric studies consist in the fact that, apart from the usual difficulties in the creation of opticomechanical systems of the required dimensions and of sufficient accuracy and rigidity, limitations of overall size and weight arise here. In addition, the conditions of functioning of the structure prove to be considerably different from ground conditions. A large degree of automation of the instrument's movements is also necessary.

In the course of the creation of the first model of an on-board submillimeter telescope of relatively large size some of the properties mentioned above were analyzed and were reflected in the concrete construction. The main parts of this model are described briefly below.

The considerations presented in [1] showed that the minimum diameter of the main mirror of a telescope intended for studies of discrete sources must be no less than 1 m. The conditions for the movement of the telescope on board and the technological possibilities for the fabrication of a mirror of acceptable accuracy limited the diameter of the main mirror to 1500 mm. Limitations in the longitudinal size led to the necessity of developing a two-mirror system for the telescope with a very deep (fast) main mirror. The following technological possibilities for the realization of the main mirror were examined: a) a glass mirror of the searchlight type with a metallized front surface; b) a metallic replica from a glass mirror; c) a metallic replica from a plastic matrix obtained by centrifugal casting; d) a metallic mirror machined out of a forging from a template on a rotary lathe; e) an optical mirror polished from a glass or pyroceramic blank.

Of the variants listed above, variants d and e could not be realized in a short period since special equipment was required. Nevertheless these variants appear to be very promising.

Variant a has the drawback that the mirror obtained is rather massive and, in addition, its inner surface apparently is not subject to exact working and consequently requires additional finishing. A very rigid mounting is required for the mirror in this variant.*

Variants b and c appeared to be the simplest in light of the fact that the technology for obtaining galvanic replicas has been developed rather fully in recent times. However, the

---

* This variant was later worked out in the details and put into reality. The results of the development will be published elsewhere.

technology for obtaining exact plastic matrices of large diameter by the method of centrifugal casting needs refinement. Therefore the variant of a metallic replica from a glass matrix was chosen as the primary variant for the draft development of the construction of a telescope. This variant had under our conditions the advantage of relative ease of fabrication, low weight, and the possibility of making multiple specimens. The drawbacks of the variant include the relatively low accuracy of the working surface, determined not only by the errors in the matrix itself but also by defects produced in taking the replica from the matirx, and deformations during temperature changes.*

The use of a relatively thin metallic replica (the thickness is determined by the technology of galvanic coating and does not exceed 6-8 mm) required the development of a rigid and light framework, also used for the mounting of the secondary hyperbolic mirror. The small field of view of the telescope (no more than 30 min) imposes a relatively strict requirement on the accuracy in aiming the optical axis (no worse than ±2 min), which excludes the rigid fastening of the mirror system to the craft. Thus, the mirror system of the telescope must have a supporting-rotating mechanism providing for its rotation about two mutually perpendicular axes.

As an analysis of the problem of telescope sensitivty [1] showed, receiving elements cooled to the temperature of liquid heliun must be used for the detection and study of discrete sources. The problem of providing relatively prolonged operation leads to the necessity of developing passive cooling systems of sufficiently large size and weight (on-board cryostats) [3]. Thus, the supporting-rotating mechanism of the telescope must carry not only the mirror but also the cryogenic system.

Here it is necessary to provide the capacity for: a) rotating the cryostat itself to an angle of 90° so that during the installation of the telescope in its various positions the longitudinal axis of the cryostat maintains a vertical position in space, which sharply reduces the helium evaporation rate; b) turning the entire supporting-rotating mechanism for servicing the cryostat in the period of preparation for the experiment; c) setting the supporting-rotating mechanism after all the preparatory operations; d) securing the cryostat and the axes of the supporting-rotating mechanism on the active section of flight; e) removing the securing devices and placing the cryostat in the fixed working position in which its longitudinal axis is parallel to the optical axis of the mirror system.

The supporting-rotating mechanism must be capable of operating under conditions of varying temperatures and a deep vacuum, which imposes rather rigid demands on the mechanical parts. Weightlessness, generally speaking, eases the demands on the weight deformations, although the dynamic loads on the active section of flight, conversely, require increased structural strength.

Steel and alloys of aluminum and titanium were considered as the construction materials. It was recognized as desirable to develop and build a laboratory model of the structure on which the necessary technological coordinations could be carried out.

The model of the mirror framework was made of titanium alloy to reduce the weight. In order to avoid complicated mechanical machining the supporting framework of the mirror was developed in the form of a welded construction of tubes with demountable fastenings to the trusses of the supporting-rotating mechanism, also built by the method of welding from steel sections and tubes. Let us describe briefly the main elements of the model's structure.

---

*Questions of the use of relatively inaccurate mirrors in a submillimeter telescope are elucidated in [2].

The supporting-rotating mechanism includes the following basic units: the framework of the main parabolic mirror with the cryostat suspension; a suspension for the secondary hyperbolic mirror; a two-axis Cardan suspension; the drive mechanism and the securing devices.

The framework of the main mirror is built in the form of an all-welded, three-dimensional, radially symmetrical, rod system of titanium thin-walled tubes. It contains 16 radial flat girders, flat chord girders, and half-braces in the upper and lower panels. Flanges which serve as the bearing points of the framework are welded to the lower belts of four chord girders which are placed at a 45° angle to the axes of rotation. Bushings with internal threading, intended for fastening the paraboloidal replica to the framework by means of threaded adjusting pins, are welded to the peripheral ring.

For technical reasons the points of intersection of the belts of the radial and inner chord girders (16 points) are connected on balls which are made hollow for lightness. Two rod brackets which serve as the cryostat suspension are fastened to the lower (back) side of the framework. Each bracket consists of a six-rod pyramid with a trapezoidal base. The housings of the journal bearings of the cryostat suspension are fastened to the tops of the pyramids. The housings of the bearings are equipped with detachable covers for convenience in mounting.

The main mirror, 1500 mm in diameter with a focal distance of 640 mm, consists of a paraboloidal shell (the replica) whose thickness varies smoothly in the radial direction from 1.5 mm in the central part to 3.5 mm at the edges.* The shell was made by the galvanic method with the successive depositing of layers of nickel—copper—nickel.

The assumed accuracy of the matrix used to make the paraboloidal replica was ±0.05 mm. The maintenance of this accuracy for the reflecting surface of the replica can be achieved through the appropriate regulation of the threaded pins by which the replica is fastened to the framework. The threaded pins are located in the bushings of the peripheral ring of the framework mentioned above and are fixed with lock nuts. The threaded pins are fastened to the shell (replica) with flanges containing ball bushings corresponding to the ball heads of the threaded pins; the flanges are fastened to the shell with four flat-headed screws. Thus, the threaded adjusting pins are fastened to the shell with swivel joints.

The mounting of the parabolic shell of the mirror to the framework and its preliminary adjustment were carried out in a special building frame using a rotating bar and a pointer indicator fastened to it. The final testing of the mirror shape was accomplished by taking aberrograms on an optical bench — an aberrograph. The quality of the mirror shape can also be checked by studying the focal spot with the mirror aimed at a point source.

The suspension for the secondary mirror consists of a four-rod three-dimensional frame in the form of a regular truncated pyramid with branching of the lower ends of the rods into two symmetrical rods. Thus, the upper (small) base of the pyramid contains four rods and the lower (large) base contains eight rods. The suspension rods are fastened to the framework of the main mirror at eight points on the peripheral ring distributed symmetrically relative to the axis of rotation of the mirror. The units for fastening the suspension to the framework allow for adjustment of the suspension by changing the length of the rods. The upper base of the suspension pyramid is made in the form of a cross consisting of a one-piece milled element of Duralumin. There are four adjusting screws on the cross for the transverse adjustment of the secondary hyperbolic mirror (diameter 300 mm) by turning its mounting on ball joints. In this case the transverse displacements are accompanied by tilting of the secondary mirror

---

*Questions of the use of relatively inaccurate mirrors in a submillimeter telescope are elucidated in [2].

relative to the main mirror. There is also the capacity for longitudinal displacement of the secondary mirror.

The Cardan suspension is designed to receive the loads from the dead weight of the mirror system and the cryostat and to make possible limited rotations (within limits of ±7°) of the mirror about the two mutually perpendicular axes, as well as the rotation of the mirror system through a 180° angle which is necessary when servicing the cryostat. The drive mechanisms are located on the suspension. The suspension consists of the following units: two mirror girders with journal bearings and the frame of the Cardan joint containing the supporting journals and a supporting girder.

The two mirror girders transform the two supporting points (journals) on the supporting rotating mechanism into four supporting points, which has considerable importance for maintaining the correct mirror shape [4]. Structurally the girder consists of a beam with a variable closed cross section of rectangular shape. The frame of the journal bearings is located in the central widened part of the girder.

The frame of the Cardan joint in plan consists of an elliptical welded frame with a shaped transverse bridge. The cross sections of the frame elements are closed with a rectangular shape. Diaphragms are mounted on the flat sections to increase the critical strength. On the major axis of the frame of the Cardan joint (through its ends) are welded the supporting journals whose axes coincide with the axis of the frame, which is arbitrarily taken as parallel to the lower plane of the frame. These journals are machined on a lathe in a single assembly and therefore must be coaxial.

On the minor axis of the Cardan joint frame are located the other two supporting journals, made in the form of detachable units, which are fastened with bolts to the upper surface of the Cardan joint frame. To assure that these journals are coaxial they are mounted on the frame with the help of a parallel matching apparatus. The inner ends of the journals, located near the center of the Cardan joint frame, are equipped with spring latches which serve to clamp the neck of the cryostat in the determined location in the working position.

The supporting girder serves as the foundation for the entire supporting-rotating mechanism by means of which the telescope is fastened on board. The supporting journals of the girder have spherical ends so that the frames of the journal bearings, which have corresponding spherical bushings, can be fastened to the craft on supporting platforms which do not require high accuracy. One of the ball journals ends in a cylindrical bushing with a diametrical channel designed to connect with the drive shaft of a worm reducer for manual turning (needed for servicing).

The drive mechanisms for both axes of the telescope contain membrane couplings with no free play, allowing for slight misalignments of the connecting axes. The transverse displacement of the axes allowed by these couplings is removed by half-ring gaskets under the flanges fastening the motors to the frame of the journal bearing; the gaskets convert the transverse displacement of the axes into an angular displacement.

The mechanism for manual turning of the telescope through 180° includes a worm reducer with a gear ratio of 1:15, a control wheel or crank with a long countershaft, a clamp to hold the support girder in the working position, and a crank with a long countershaft for the clamp.

The clamp for holding the support girder in the working position consists of a pin with a conical end located within the cylindrical part of a socket fastened to the frame of the bearing of the supporting girder. The conical part of the pin enters a corresponding opening in the supporting journal of the girder. The pin is subjected to a longitudinal compression force by a cylindrical spring, preventing the pin from coming out of the recess in the supporting journal. The upper part of the socket within which the locking pin is located ends in a conical funnel

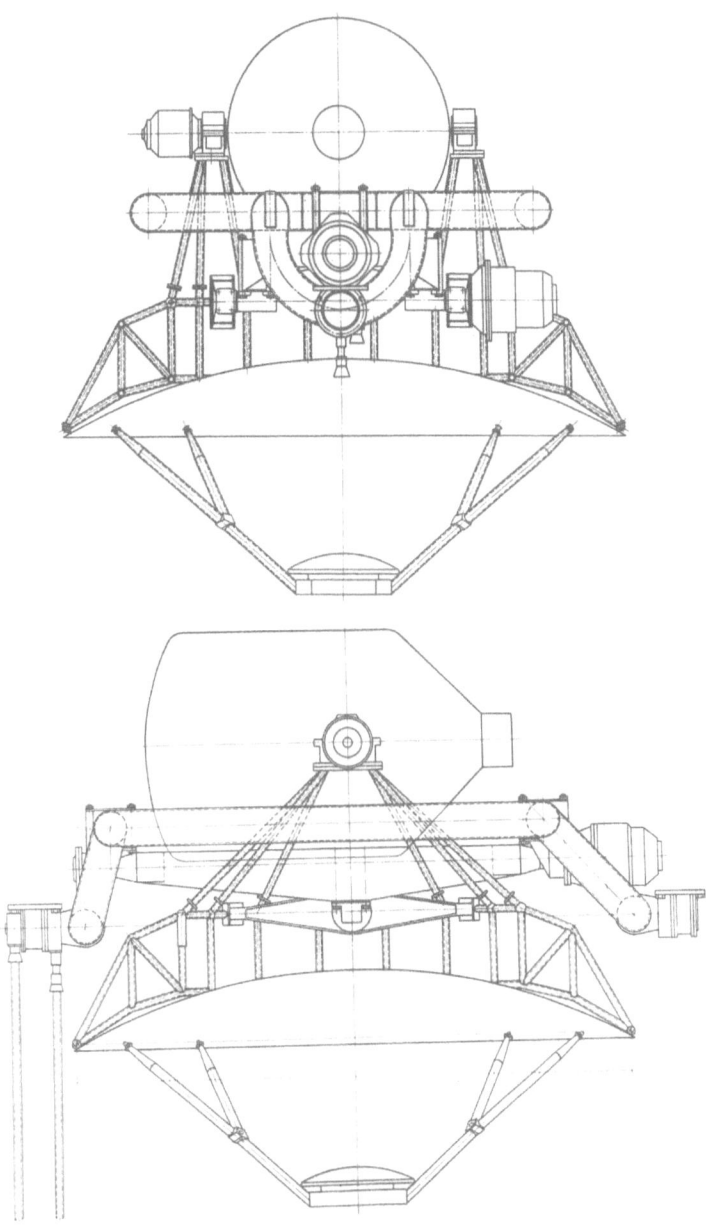

Fig. 1. General views of construction of telescope model (cryostat in transport position).

Fig. 2.   Photograph of model on transporting frame
(cryostat in working position).

which also acts as an opening for the worm reducer of the manual drive mechanism. Using
a crank with a long shaft the locking pin can be turned about its axis to a position where it
is freed from support by its cross pin in the body of the socket and is withdrawn from the recess in the supporting bearing by a manual force on the shaft overcoming the force of the
spring. The freed supporting girder can be turned by the hand drive mechanism.

General views of the structure of the telescope model developed and fabricated at the
Institute of Physics are presented in Fig. 1. In the photograph (Fig. 2) the model is shown
with the cryostat in the working position. The supporting-rotating mechanism of a standard
searchlight mount is used for transporting and turning the model as a whole; the telescope
model is fastened with a special adapter frame.

In conclusion, we note that the preliminary development of the mechanical construction
of the telescope demonstrated the possibility of its placement and functioning under on-board
conditions. The use of a main mirror of increased accuracy, which allows the use of single-
mode receivers, appears to be very promising. Since an increase in accuracy can be achieved
only by optical methods, it is appropriate to consider the variant mentioned above using a
pyroceramic or glass mirror finished with the accuracy attainable in an optical shop.

The use of all-welded or partly-welded constructions, while facilitating the building of
the model, creates in addition the danger of distortions if the annealing is not deep enough.
Mechanical machining of the construction would be preferable in the presence of large-scale
equipment and casting. The development of such a variant must be connected with the specific
production possibilities.

LITERATURE CITED

1.   A. E. Salomonovich, Tr. FIAN, 77, 33 (1974).
2.   A. S. Khaikin, Tr. FIAN, 77, 56 (1974).
3.   A. B. Fradkov and V. F. Troitskii, Tr. FIAN, 77, 85 (1974).
4.   P. D. Kalachev, Izv. Vuz. Mashinostr., No. 12, 9 (1963).

AN ON-BOARD SUBMILLIMETER SPECTRORADIOMETER

A. A. Kobzev, V. I. Lapshin, V. F. Troitskii, and A. S. Khaikin

Studies of the spectrum of the sun's radiation in the submillimeter wavelength range are of considerable interest for the determination of the physical conditions in the lowest layers of its atmosphere (temperature, active regions) [1] and for the detection of highly ionized atoms of heavy elements in the solar corona [2]. The sun, as the most intense source of extraterrestrial submillimeter radiation, is at the same time an appropriate subject for tests of spectroradiometers of this range.

Although certain measurements of the sun's submillimeter radiation are possible under ground conditions (especially in the long-wave part of the range), extra-atmospheric studies prove to be the most informative — those conducted on balloons [3, 4] and on geophysical rockets, i.e., at altitudes where the absorption in the water vapor and other gases of the earth's atmosphere is negligibly small.

During the development of the first examples of on-board submillimeter spectroradiometers intended for extra-atmospheric studies we developed and tested a solar submillimeter spectroradiometer (SMS) for the region of 300-1000 μ. A brief description of the setup of the main units of the instrument and the results of its studies are presented below.

The spectroradiometer was designed for installation in the instrument bay of a Vertikal' geophysical rocket, which determined its overall size and several other parameters.

The field of view of the optical system of the instrument had to be on the order of 40' since it was proposed to receive the radiation of the solar disk as a whole. The experimental conditions required the mobility of the optical axis in the limits of $\pm 30°$ in azimuth and from -5 to $+20°$ in elevation. In order to hold the radiometer axis in the direction toward the center of the sun's disk an automatic aiming system for the two axes with an accuracy of $\sim 10'$ was required. Although the expected brightness temperature of the sun in the region of the spectrum measured (300-1000 μ) is rather high ($\sim 5 \cdot 10^3 °K$), a receiver cooled to the temperature of liquid helium proves to be necessary for spectral measurements. As the spectral instrument it was decided to use a Fourier spectrometer, making it possible to record several interferograms with satisfactory resolution in the relatively brief time of the experiment (~ 10 min).

An optical-kinematic diagram of the SMS is presented in Fig. 1. The sun's image is obtained with a Cassegrain telescope objective. Its focal distance is 976 mm, the main mirror is 160 mm in diameter, the secondary is 80 mm, and length is 75 mm. Spherical mirrors were used in the telescope objective and therefore the sun's image was increased by spherical aberration from 7 to ~ 12 mm. The calculation and the characteristics of the telescope objective are described in detail in [5]. The diameter of the receiver entrance window is 11 mm.

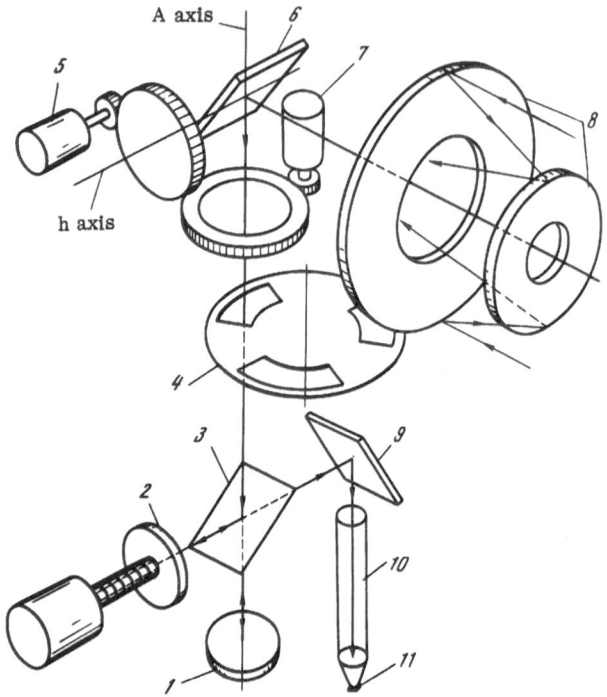

Fig. 1. Optical-kinematic diagram of spectroradiometer. 1) Fixed
mirror; 2) movable mirror; 3) light divider; 4) modulator; 5) motor
for aiming in elevation (h); 6) rotating mirror, rotation angle h/2;
7) motor for aiming in azimuth (A); 8) telescope; 9) 45° mirror; 10)
light pipe (inside cryostat); 11) receiver.

The necessity of providing for mobility of the axis of the telescope objective led to some
complication of the system. It was possible to keep the entire spectroradiometer stationary
relative to the instrument bay except for the telescope objective; rotation of the latter in azimuth A takes place about the transform of the light pipe axis, while the rotation in elevation
h is compensated by rotation of the plane rotating mirror through an angle h/2.

The kinematics of the rotation mechanism is shown in Fig. 1. The automatic aiming
system of the relay type uses differential solar pickups. When the axis of the telescope objective deviates from the direction toward the center of the solar disk the illumination of a pair
of silicon photoelements, sensitive to visible light, is unbalanced and the signals produced
in them after amplification act on the motors of the azimuth and elevation drives, which rotate
the axis of the telescope objective to the direction toward the center of the solar disk. The
aiming accuracy is ±8'.

After reflection from the plane compensating mirror the radiation is modulated by a
disk modulator (modulation frequency 180 Hz) and enters a Michelson interferometer, the
movable mirror of which is set into motion by a direct current electric motor with a stabilized
rotation rate. The mirror moves back and forth, covering a path of 20 mm in 00 sec. A divid-

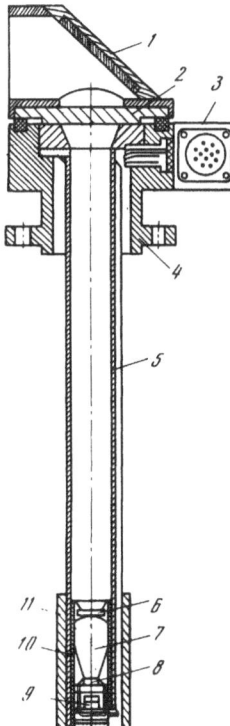

Fig. 2. Construction of receiver. 1) Deflecting
mirror; 2) window (Teflon); 3) output socket; 4)
seal; 5) light pipe; 6) filter (black paper); 7)
immersion lens (germanium); 8) n-InSb crystal;
9) resistance thermometer; 10) cup (Teflon); 11)
shield (brass).

er of polyethylene terephthalate (PETP) with a thickness of 50 μ, which is efficient in the region of 300-1000 μ, is used in the interferometer. The divider also primarily determines the
working spectral range of the instrument. The true resolving power of the interferometer at
the wavelength $\lambda \approx 500\ \mu$ is ~40.

The radiation emerging from the interferometer, reflecting from a stationary deflecting
mirror, falls on the entrance opening of the light pipe, where an image of the sun is formed,
and then passes along the light pipe to the receiving element.

A receiving element of n-InSb cooled by liquid helium is used in the spectroradiometer.
The construction of the receiver [6] is shown in Fig. 2. The photoresistor $4 \times 4 \times 1$ mm in
size with an immersion lens of germanium is placed at the lower end of a polished cylindrical
light pipe 11 mm in diameter made of stainless steel. A resistance thermometer is also
placed here to monitor the temperature of the receiving element. The upper part of the light
pipe is closed by a Teflon window and has a flange for mounting on the neck of the cryostat.
The deflecting mirror and a socket containing the outputs from the receiving element and the
thermometer are located on the flange.

The cryostat (Fig. 3) was developed specially for the radiometer. Because of the conditions of the experiment the placement of the helium-filled cryostat in the instrument bay can
take place not less than 5-6 h before the launch. Therefore although the duration of the experiment itself does not exceed 10 min, the cryostat must keep the liquid helium at a temperature

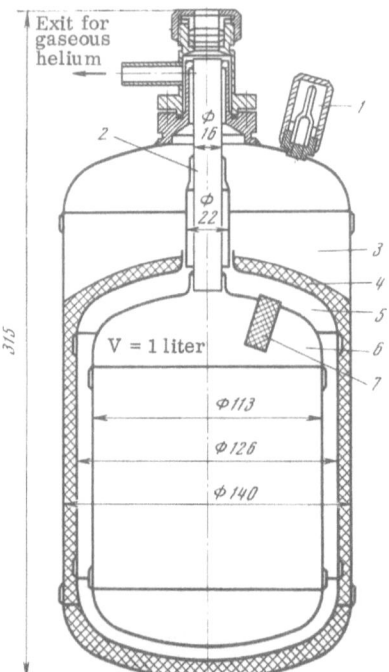

Fig. 3. Nitrogenless KKR-2 helium cryostat.
1) Evacuation unit; 2) neck (tube of stainless steel); 3) outer casing (copper); 4) shield-vacuum insulation; 5) shield (copper); 6) helium volume; 7) adsorbent.

Fig. 4. General view of cryostat and receiver.

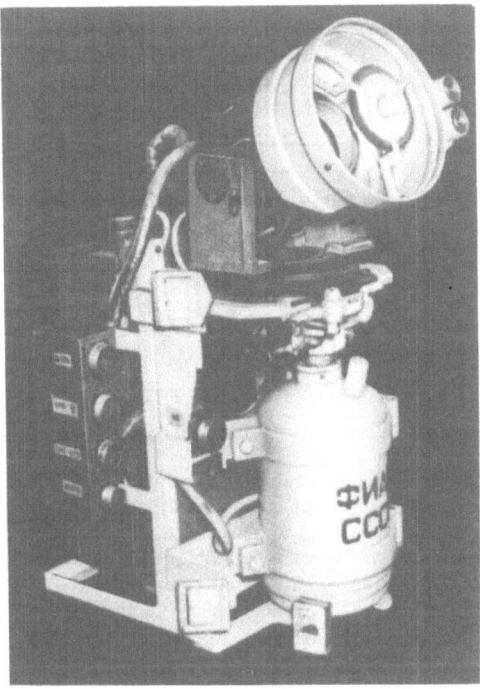

Fig. 5. General view of spectroradiometer.

of 4.2°K for at least 10 h. The KKR-2 cryostat does not have nitrogen cooling; to reduce the inflow of heat a multilayer shield-vacuum heat insulator of aluminized PETP film is sandwiched between the intermediate shield and the outer vacuum casing. The volume of liquid helium in the cryostat is 1 liter and the duration of its retention under ground conditions is from 10 to 20 h depending on the quality of the vacuum between the shields. The height of the cryostat is such that the receiving element is located near its bottom.

A general view of the cryostat and receiver is shown in Fig. 4. An automatic pressure regulator which maintains atmospheric pressure in the cryostat during the experiment is also mounted on the flange of the receiver.

A general view of the spectroradiometer is presented in Fig. 5. An electrical block diagram of the instrument is given in Fig. 6. All the electronic units are transistorized except for the bulb of the photopickup for the reference signal.

The signal amplifier together with the synchronous detector and direct current amplifier have a maximum through amplification of $\sim 2.2 \cdot 10^5$ with an input resistance $R_{in} \simeq 3$ kΩ. The internal noise of the input amplifier is $\sim 10^{-7}$ V at a modulation frequency of 180 Hz and a transmission band (to the detector) of ± 90 Hz. The time constant of the synchronous detector is 50 msec and the output voltage of the DCA is from 0 to 7 V at $R_{out} \simeq 500$ Ω, which is determined by the parameters of the telemetry system. The through amplitude characteristic of the amplifier is linear up to an output voltage of $\simeq 3$ V; at voltages of more than 3 V it is logarithmic with a coefficient of compression of the dynamic range of 8.

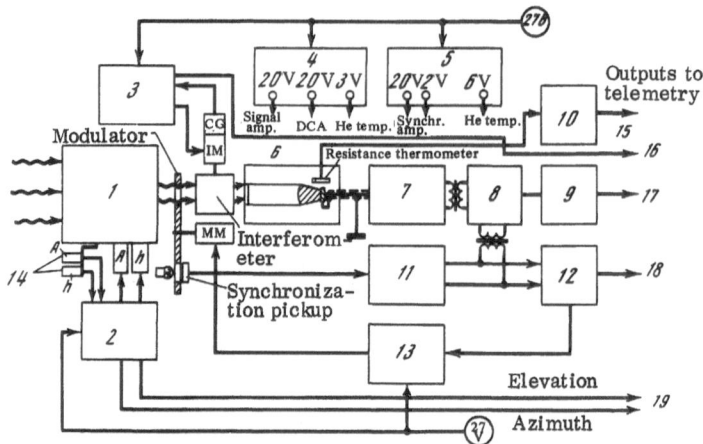

Fig. 6. Electrical block diagram of instrument. 1) Telescope on rotating mechanism; 2) servo mechanism; 3) interferometer velocity stabilizer; 4) power transformer 1; 5) power transformer 2; 6) cryostat; 7)
signal amplifier; 8) synchronous detector; 9) direct current amplifier;
10) helium thermometer amplifier; 11) synchronization amplifier; 12)
modulator velocity meter; 13) modulator velocity stabilizer; 14) solar
pickups (on telescope); 15) helium temperature; 16) direction of interferometer travel; 17) signal; 18) modulator velocity; 19) control for operation of aiming motors.

Besides the amplifier the electronics assembly contains stabilizers for the rotation rates
of the electric motors of the interferometer and the modulator. For the measurement of the
helium temperature there is a balanced direct current amplifier whose arm includes a resistance thermometer, and the output voltage varies from 0 to 3 V with a change in temperature
from 5.7 to 1.3°K. All the electronics units of the instrument are supplied from an on-board
voltage supply at 27 V through two voltage transformers of direct current to direct current.
The transformers perform the following functions: 1) they stabilize changes in the voltage supply; 2) they galvanically uncouple the instrument from the power supply and suppress interference coming from the supply; 3) they generate voltages differing from 27 V and required for
the supply of different units. The electronics of the aiming system is placed in a separate
assembly.

After assembly and adjustment the spectroradiometer underwent laboratory tests. The
radiation spectrum of a PRK-4 lamp obtained on this instrument during the tests is shown in
Fig. 7. The thickness of the layer of air containing normal humidity was about 70 cm. The
aiming system was aligned with respect to the sun on the mount of the solar telescope of the
P. K. Shternberg State Astronomical Institute. The spectroradiometer was mounted on a platform imitating external disturbances. The accuracy of the tracking on the sun in azimuth and
position angle was no worse than ±8'. Locking onto the sun occurred in the ranges of angles
of ±15° in h and ±10° in A.

A flight experiment occurred on August 20, 1971, with the elevation of the sun above the
horizon h = 12° and the azimuth A ≃ 0° relative to the instrument axis. The Vertikal' rocket
rose to an altitude of 463 km. The spectroradiometer was turned on according to the program

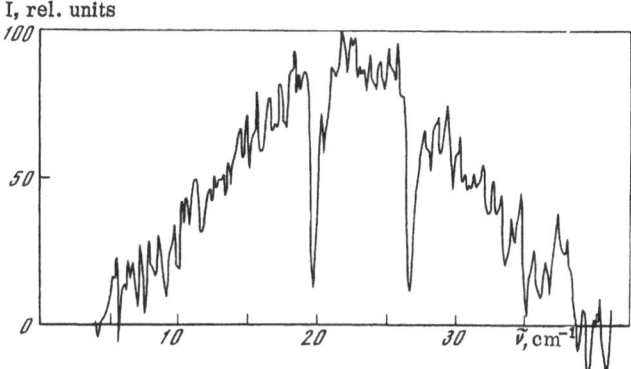

Fig. 7. Reproduced spectrum of radiation of a PRK-4 mercury lamp. Interferogram recorded during tests of the instrument in the laboratory; thickness of the air layer containing normal humidity was about 70 cm.

after the engines shut down and operated for 10 min. The telemetry recordings indicate the normal operation of all the units of the instrument except for the solar pickup for aiming in elevation. The aiming took place only in azimuth. The telescope was almost stationary in elevation and its axis was aimed somewhat below the horizon. On the recordings of the output voltage of the signal amplifier at the times of the programmed closing of the amplifier input one can see steps corresponding to the difference between the brightness temperatures of the modulator disk (~ 300°K) and of the upper layers of the earth's atmosphere (~250°K). This indicates normal sensitivity of the entire receiver-amplifier circuit in flight.

The temperature variations of individual units of the instrument in the course of the experiment are shown in Fig. 8. The temperature pickups were located on the back side of the main mirror, in the window of the modulator, and in the electronics assembly. As is seen, the temperatures varied little in the course of the flight. The temperature of the helium in the cryostat changed sharply only during acceleration and thanks to the pressure regulator it was kept practically constant on the entire passive section of flight.

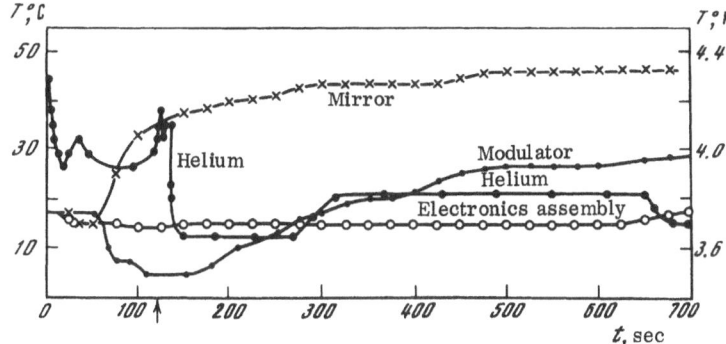

Fig. 8. Temperature changes of instrument units in flight.

The data on the functioning of the SMS instrument were used in the subsequent development of on-board submillimeter telescopes.

The authors thank A. E. Salomonovich and A. B. Fradkov for attention and assistance in performing the work, A. E. Bor-Ramenskii and I. S. Rybkin for the development of the aiming system, and S. N. Churilov for help in the tests and in the preparation for conducting the experiment.

LITERATURE CITED

1. A. E. Salomonovich, Usp. Fiz. Nauk, 99, 417 (1969).
2. I. L. Beigman, L. A. Vainshtein, and R. A. Syunyaev, Usp. Fiz. Nauk, 101, 607 (1969).
3. J. Gay, J. Lequeux, J. P. Verdet, P. Turon-Lacarrier, M. Bardet, J. Roucher, and Y. Zeau, Astrophys. Lett., 2, 169 (1969).
4. P. Stettler, F. K. Kneubühl, and E. A. Müller, Astron. Astrophys., 20, 309 (1972).
5. V. S. Ivleva, A. A. Kobzev, V. I. Lapshin, A. E. Salomonovich, V. I. Selyanina, V. F. Troitskii, A. B. Fradkov, and A. S. Khaikin, Preprint FIAN No. 12 (1971).
6. A. S. Khaikin, Tr. FIAN, 77, 56 (1974).

POLARIZING DEVICES FOR THE SUBMILLIMETER RANGE

V. I. Lapshin

As is known, measurements of the polarization characteristics of radiation which is received in parallel with the study of its spectral composition provide valuable information of the phenomena and processes in which this radiation takes part. The interest in various devices which permit the analysis and measurement of the state of polarization of radiation in a wide region of the spectrum is understandable in this connection.

The development of measuring techniques for the submillimeter range (wavelengths λ = 0.05-2.0 mm), stimulated in particular by the demands of numerous astrophysical and aeronomical problems [1, 2], leads to the necessity of developing various polarizing elements and devices for this range.

In the present report we examine several new submillimeter-range devices intended for use in polarimetric and spectroscopic research.

In Section 1 we examine lattice polarizers of a special type, which are the basic elements for twin-wave polarization interferometers examined in Sections 2 and 3. Section 3 is devoted to a description of several new interferometer systems.

An increase in the signal-to-noise ratio through the use of the method of Fourier spectroscopy [3-5] is very important in the submillimeter range where the sensitivity is often limited by the receiving devices. The potential capabilities of twin-beam interferometers could not be realized until recently mainly because of the absence of sufficiently efficient and broad-band dividers. The use of lattice polarizers, which have good polarization characteristics in a broad region of the submillimeter spectrum, allows one to propose a number of new interferometer systems in which the polarization method of dividing the light beam into coherent components is used.

The method of Fourier spectroscopy also attracts attention in connection with the possibilities for its use in polarimetric measurements [6]. The development of polarization interferometers has great importance from this point of view, since here it becomes possible in practice to achieve in one instrument the advantages inherent to spectral and polarimetric instruments. It must be noted that despite the obvious advantages of the study of the polarization-spectral characteristics of radiation, at present we cannot point to a single sufficiently complete report where such measurements have been made in the submillimeter region of the spectrum. Evidently, this must be explained entirely by the absence or insufficient development of the necessary technology.

1. METAL-FILM POLARIZERS FOR THE SUBMILLIMETER RANGE*

The main purpose of polarizers is the isolation of electromagnetic radiation having a certain direction of the wave vector of the electromagnetic field. Various one-dimensional arrays of filaments, belts, and ribbons are widely used as polarizers in the submillimeter range. Lattice polarizers which consist of one-dimensional grids, composed of parallel thin (the transverse sizes are much less than the wavelength) metal conductors stretched on a special mount at equal distances from one another, are well known [7-10]. The spatial period g (the distance between adjacent conductors) of such a grid essentially determines its optical properties (the reflection and transmission coefficients) and is taken as small in comparison with the wavelength to obtain good polarization characteristics.

The characteristic curves of such polarizers made in the form of one-dimensional conducting grids [7-10] are satisfactory in the middle and long-wave sections of the range ($\lambda >$ 300 μ), so that for this region of the spectrum it is comparatively easy to satisfy the condition $g/\lambda \ll 1$, which determines the efficiency of the polarizer (see [7], for example). But for the short-wave part of the range the satisfying of this condition encounters serious technical difficulties, since the fabrication of a filament grid with a period g small enough for wavelengths $\lambda < 300~\mu$ becomes very complicated. Therefore, despite the wide use which filament elements have received in submillimeter interferometry [9], attempts have repeatedly been made to create lattice polarizers of other types, particularly [11] those which in a later discussion we will call metal-film type polarizers. They consist of rather thin (the thickness is less than the wavelength) dielectric films (polyethylene, polyethylene terephthalate) with metal coatings in the form of equidistant parallel strips deposited on them.

The main advantage of such polarizers is the possibility in principle of preparing by rather simple means grids with periods on the order of tens of microns, which considerably expands the region of their applicability. The use of more complicated techniques [12, 13] allows one to prepare metal-film polarizers (MFP) with periods of ~1 μ which display high polarizing ability even in the infrared range of the spectrum. In this case, however, an additional element appears in comparison with filament strcutures — the substrate, whose characteristics can, generally speaking, have a considerable influence on the properties of the MFP and therefore must be studied and allowed for in their preparation.

In our case polyethylene terephthalate (Mylar) PÉTF-DA films 5 and 18 μ thick with an aluminum coating 0.05-0.10 μ thick served as the material for the preparation of the MFP. The method of cutting the MFP is similar to that described in [11]: the film was fastened and stretched on a special cylindrical mounting and slits were made in the metal coating on a lathe with automatic longitudinal feeding without disturbing the dielectric substrate. However, in contrast to [11] where emery paper was used as the cutting element, in our work a cutting tool made from a razor blade and fastened to a mounting which provides a constant pressure on the cutting edge proved more suitable. One can obtain MFPs with different periods by regulating the feeding rate. In our case the minimum step in the longitudinal feeding of the lathe was 18 μ, which assured that the condition $g/\lambda \ll 1$ is satisfied in almost the entire submillimeter region.

For a preliminary estimate of the effect of the dielectric substrate we measured the transmission spectra of Mylar films (without the metal coating) in the spectral range of $\lambda =$ 40-1000 μ on a Hitachi FIS-21 spectrometer.

---

* The principal results of this section were obtained jointly with V. A. Vagin, a graduate student at the Moscow Physicotechnical Institute. Additional details of a technical and methodological nature are presented in his thesis prepared at the Institute of Physics in 1972.

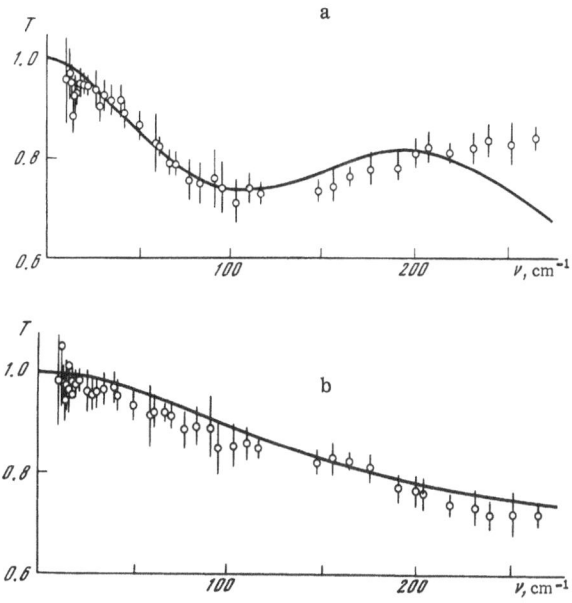

Fig. 1. Transmission of polyethylene terephthalate film 18 μ
thick (a) and 5 μ thick (b).

The experimental values obtained for the transmission coefficients of films 5 and 18 μ
thick as a function of the wave number ($\nu = 1/\lambda$) are shown by vertical segments in Fig. 1.
At the same time calculations were made of the transmission of the films, allowing for Fresnel
reflections from the film boundaries, multibeam interference within it, and radiation losses
due to absorption [14]. The curves obtained are presented in Fig. 1 with solid lines. Here the
index of refraction of Mylar was taken as constant in the region of the spectrum under consideration and equal to n = 1.6; the absorption coefficient was taken as k = 0.05 [15, 16]. As is
seen from the figures, the measured spectral dependences of the transmission coefficients of
Mylar films agree rather well with theoretically calculated dependences for the values of the
parameters n and k adopted. The studies conducted showed that polyethylene terephthalate
films 5 and 18 μ thick are transparent enough in the spectral region of 40-1000 μ to allow their
use as substrates for MFPs. It should also be noted that since interference in a film 18 μ thick
leads to a certain reduction in its transmission in the region of $\lambda < 100$ μ, films 5 μ thick are
evidently preferable for work in the short-wave part of the submillimeter range at least.

Measurements of the characteristic curves of the MFP prepared were also made on an
FIS-21 spectrometer by a method, described in detail in [8, 12], which allows one to take into
account the polarization characteristics of the spectrometer itself. The spectral dependences
of the transmission coefficients T_E and T_H (the polarizations with electric vectors parallel and
perpendicular to the lines, respectively) of the MFPs were measured. The results obtained
are shown by vertical segments in Fig. 2 for films 5 and 18 μ thick. As should be expected,
the highest transmission (90-95%) for H-polarization is observed at low frequencies. There
is a second transmission maximum for this polarization in the region of g/λ $\approx$ 0.3 for the
18 μ film. In the case of the 5 μ film a monotonic decrease in the transmission of the H-

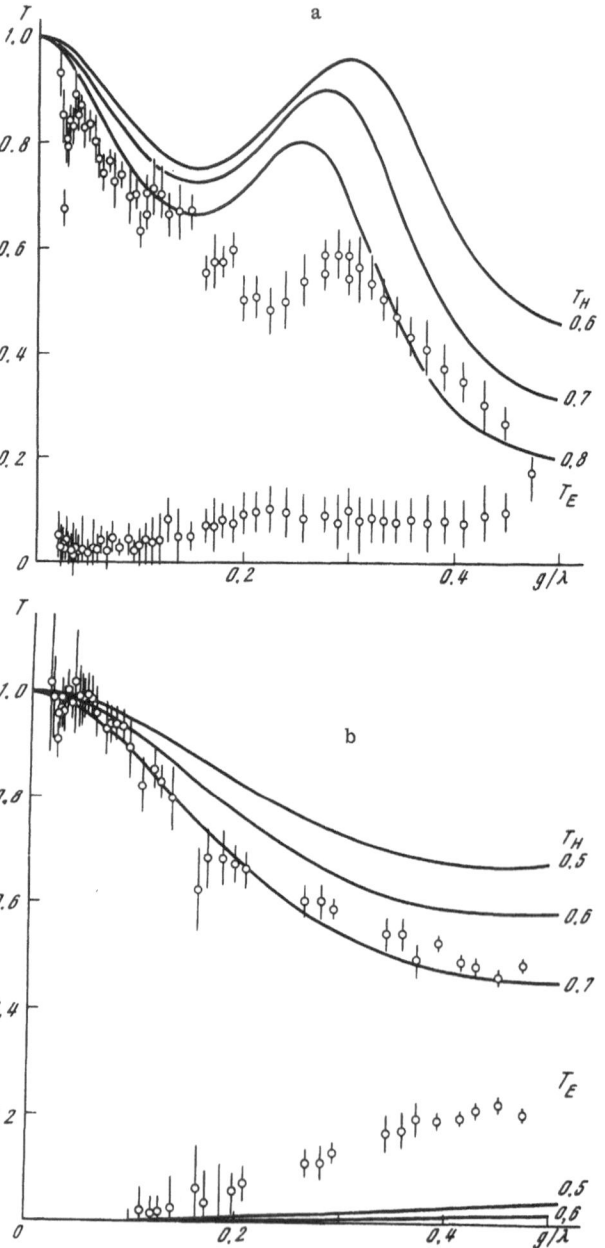

Fig. 2. Transmission of waves with E- and H-polarization by metal-
film polarizers 18 μ thick (a) and 5 μ thick (b).

polarization occurs with an increase in frequency. The frequency dependence of the T_E coefficients is described by a monotonic function for both films.

It was interesting to compare the measured MFP characteristics with theoretical estimates. The problem of the diffraction of a plane electromagnetic wave on a metal lattice has been examined in [17-19]. The case of $g/\lambda \ll 1$ for a lattice of filaments of round cross section is calculated most accurately. With the incidence of a plane wave only the reflected and transmitted waves are examined in this case since the diffracted waves have a surface nature and affect only the distribution of the field near the conductors. When $g/\lambda \ll 1$ this field can be considered as quasi-steady, which allows one to determine the boundary conditions and to calculate the reflection and transmission coefficients for waves of different polarizations. Such a treatment is applicable to existing filament polarizers in the region of $\lambda > 120~\mu$.

A method for solving the problem of the diffraction of electromagnetic waves on an infinite lattice of thin ribbon bands for arbitrary values of the lattice parameters is proposed in [18, 19] (lattices having a dielectric substrate are considered in the latter report). For the unknown amplitudes of the reflected and transmitted waves one obtains an infinite system of equations whose solution permits one to obtain both approximate and exact (in the form of infinite series) expressions for the transmission coefficients.

We calculated the T_E and T_H coefficients of the MFPs from the equations of [18, 19] using a Nairi-2 computer. The programs compiled permitted the variation of different parameters of the polarizers (the filling coefficient b, equal to the ratio between the width of a metal band and the lattice period, the index of refraction, and the substrate thickness). The results of the

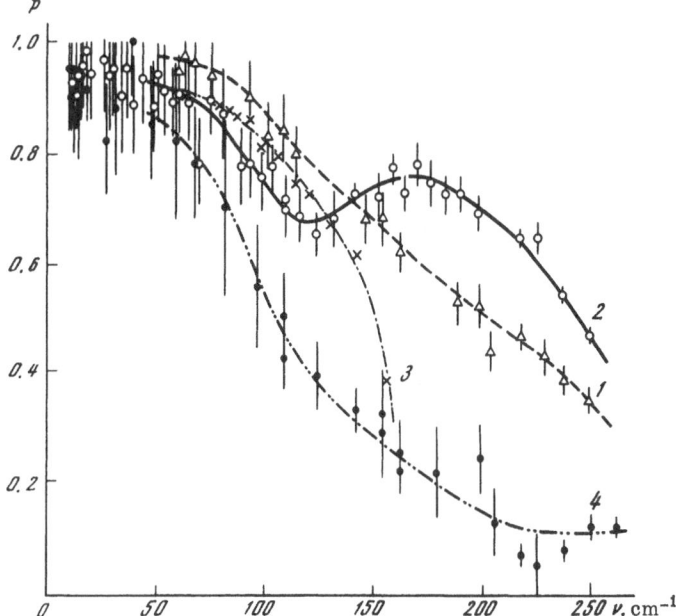

Fig. 3. Degree of polarization of polarizers of different types. 1) $5~\mu$ MFP; 2) $18~\mu$ MFP; 3) filament polarizer [8]; filament polarizer measured: $g = 40~\mu$, filament diameter $8~\mu$.

calculations are shown in Fig. 2 by solid curves for different values of the coefficient b. As seen from the figure, the calculated dependences of the T_E and T_H coefficients on g/λ for polarizers with coefficients b = 0.5-0.7 agree qualitatively with the experimental values. However, the theoretical values of the coefficients prove to be greater than the measured values. The parameter b of the polarizers which we prepared is not fixed: Its value could vary over the area of the lattice in the range from 0.5 to 0.9. The form of the $T_H (g/\lambda)$ and $T_E (g/\lambda)$ curves shows, however, that this fact cannot fully explain the quantitative disagreements observed in the region starting with $g/\lambda > 0.1$. The presence of absorption in the substrate material, which is not allowed for in the equations of [19] used, seems to us a more likely explanation for the results obtained. The allowance for absorption in the substrate considerably complicates the calculation of the MFP characteristics, however.

It also seems interesting to compare the polarization characteristics of the proposed MFPs with the data on filament polarizers, such as those presented in [8] for polarizers with a period g = 30 μ (filament diameter 8 μ) and those measured by us (g = 40 μ, diameter 8 μ). It is convenient to compare different polarizers with respect to the degree of polarization, determined [8] as $P = (T_H - T_E)/(T_H + T_E)$. Graphs of the dependence of P on ν for polarizers of different types are presented in Fig. 3 for the region of frequencies ν from 10 to 250 cm^{-1}. It is seen from the graphs that the degree of polarization of MFPs in the region of $\nu < 100$ cm^{-1} is no less than for the better filament polarizers. In the region of $\nu > 100$ cm^{-1} the MFPs have considerably better polarization characteristics.

Thus, the spectral studies made of the polarization characteristics of the MFPs developed confirmed the possibility of using them in the entire submillimeter range. The transmission coefficients calculated for the MFPs also showed the possibility of the preliminary estimate of MFP characteristics, which makes it possible to select the optimum MFP parameters from the point of view of the requirements of a specific problem.

2. POLARIZATION INTERFEROMETERS FOR THE SUBMILLIMETER RANGE

For practical purposes an important property of both filament and metal-film lattice polarizers for the submillimeter range, distinguishing them from the dichroic polaroids used, for example, in the visible region of the spectrum, is the possibility of using them in different variants of an instrument both "in transmission" and "in reflection." The passed radiation is polarized: The component of the electric vector of the wave is directed perpendicular to the lattice (H-polarization); the reflected radiation has a mutually orthogonal direction (E-polarization). When a plane polarized wave whose plane of linear polarization forms a certain angle with the direction of the filaments is incident on a grid, the passed and reflected components remain coherent. This makes it possible to use linear grids as wide-band light dividers for twin-beam interferometers, such as the Michelson type of interferometer, which have obtained wide distribution in connection with methods of Fourier spectroscopy. This feature of lattice polarizers was first pointed out in [3, 7].

It must be noted that since the light divider is the most important element of a twin-beam interferometer, determining its transmission and effective spectral range, principal attention in the development of Fourier interferometers is devoted to the analysis and study of various dividers of submillimeter radiation. The variants of twin-beam interferometers with spatial division of the wave front described in [20-22], while they permit the solution of the problem of making light dividers wide-banded and efficient, have in turn a number of fundamental drawbacks, such as the difficulty of obtaining large path differences in the system of [20, 21] or assuring stability of the uniform illumination in the system of [22]. The comparative simplicity of construction of the classic system of the Michelson interferometer, on the one hand, and the enumerated drawbacks of other twin-beam interferometer devices, on the other, have resulted in the fact that systems of twin-beam interferometers with amplitude division of the light beam have still obtained the widest distribution.

The detailed examination of light dividers of different materials (dielectric and semiconducting films, paper, etc.) carried out in [15, 16] and the examination of special constructions in the form of prisms with the total internal reflection disrupted [23] showed that the optimum efficiency of a submillimeter interferometer for a given section of the spectrum can be provided by the special selection of the parameters of the material (the index of refraction and the thickness) in [15, 16] or of the gap between prisms in [23]. However, the problem of creating a light divider which is optimal in a wide range of the submillimeter spectrum had not been solved up to now.

From this point of view the emergence of wide-band polarizers opens up new possibilities for the construction of interferometer devices for the submillimeter range which are free from the drawbacks mentioned for the systems examined. It is also important that in this case it becomes possible to record the polarization characteristics of radiation and their spectral variations, which is also impossible to accomplish using the standard means of interferometry.

There are presently descriptions in the literature [3, 11, 24-26] of several systems of polarization interferometers. The principle suggested in [3] is used in them, although the systems of construction differ. The principle of operation of a polarization interferometer for the submillimeter range can be explained as follows. The radiation studied, passed through a polarizer of submillimeter radiation, is directed at a second lattice polarizer arranged so that the plane of polarization of the beam incident on it forms an angle of 45° with the direction of the filaments. With this arrangement of the lattice polarizer the beams passing through it and reflected from it are coherent and have the same amplitudes and mutually orthogonal polarizations. The reduction of these beams into one using a system of mirrors (in this

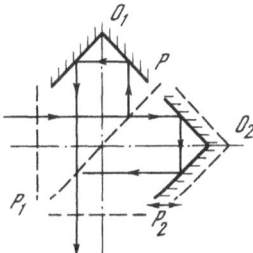

Fig. 4. Optical diagram of polarization interferometer [11]. The direction of the filaments of polarizer P_0 is parallel to, while that of polarizer P_1 forms a 45° angle with, the edges of reflectors O_1 and O_2; polarizer P rotates about the longitudinal axis with a constant angular velocity.

Fig. 5. Optical diagram of polarization interferometer [24]. The directions of the filaments of polarizers P_1 and P_2 form a 45° angle with each other.

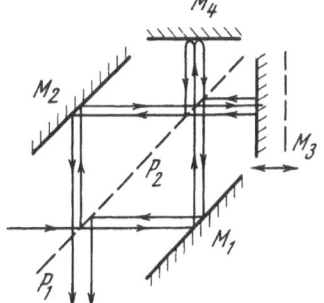

respect the overall system of the interferometer to a considerable extent retains a similarity with the arrangement of the elements in the Michelson interferometer) allows one to obtain a beam with a polarization whose form varies as a function of the phase difference of the interfering beams. The state of polarization of the emerging radiation is analyzed with another polarizer which converts the phase changes of the interfering beams into amplitude changes of the emerging beam. Diagrams of the interferometers of [11, 24] are shown in Figs. 4 and 5.

The construction differences in the systems of [11, 24] lead to a certain difference in their characteristics. In particular, as noted in [11], a positive characteristic of the system is, for example, the possibility of eliminating the constant component in the interferogram, which permits a reduction in the effect of fluctuations in the intensity of the radiation source. The system of [24], however, provides the same luminosity as that of [11] in work with polarized radiation since both polarizations of the radiation isolated by the first polarizer are used in it. With ideal polarizers the efficiency of this system for polarized radiation is equal to the efficiency of an ideal Michelson interferometer. Both systems can be used for spectro-polarimetric measurements.

Let us dwell in somewhat more detail on the latter question. The presence of polarizers of submillimeter radiation allows one in principle to conduct polarimetric measurements with any of the spectral instruments known earlier, including Michelson interferometers. In this respect the polarization interferometers, in addition to their advantages mentioned earlier, permit a decrease in the number of components needed, since the polarizers in them perform the function of light dividers also. The possibility of using the method of Fourier spectroscopy to obtain complete information on the state of polarization of the radiation incident on the interferometer was studied in a series of reports [6, 27]. It is shown that with polarizers mounted in each of the arms of a twin-beam interferometer three independent interferograms, corresponding to the three different mutual orientations of the polarizers, are needed to determine all four of the Stokes parameters [28]. The general analytical method developed in these works, using the formalism of coherence matrices and Jones matrices to describe the fluxes of polarized radiation in the interferometer, allows one to take into account the various polarization effects in Fourier spectroscopy. Thus, both the sufficiently effective means and the theoretically justified method of spectropolarimetric measurements are now worked out in submillimeter spectroscopy.

3. SOME NEW SYSTEMS FOR POLARIZATION INTERFEROMETERS

The development of lattice polarizers opens up possibilities for the construction of new interferometer systems for the submillimeter region. As an example let us examine two new modifications of polarization interferometers. The corner reflector, employed earlier in [11], is used in them. For an explanation of its operation let us consider Fig. 6.

Polarized radiation whose plane of polarization forms an angle θ with the edge of the reflector enters the corner reflector, which consists of a dihedral angle formed, for example,

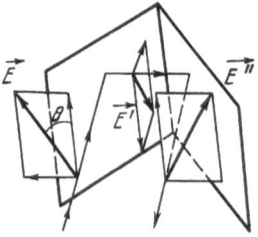

Fig. 6. Reflection of a beam of polarized radiation from a corner reflector.

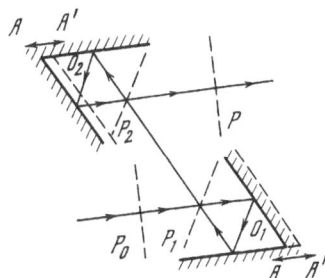

Fig. 7. Diagram of a polarization interferometer for obtaining large path differences with
small mirror displacements.

by two mirrors. Taking into account the phase change through π which occurs during the reflection of each component, it is easy to ascertain that the plane of polarization of the beam
after reflection proves to be turned around to the angle $-\theta$ relative to the edge of the reflector.
Thus, such a reflector permits one to rotate the plane of polarization of a beam through an
arbitrary angle without losses.

The diagrams of two new interferometer modifications are shown in Figs. 7 and 8. The
first of them is designed to obtain considerable path differences of the interfering beams with
the minimum movement of the interferometer mirrors. Let us examine the operation of this
system. The interferometer is formed by two corner reflectors O_1 and O_2 whose edges are
parallel and whose angles are equal to 60° and by two lattice polarizers P_1 and P_2. The polarizer P_0 is necessary to isolate the polarized component at the entrance to the interferometer
while the polarizer P is used as the analyzer of the polarization of the emerging radiation.
The direction of the filaments (the "polarization axis") of P_0 is set parallel to the edge of reflector O_1. The axis of polarizer P_1 is set at a 45° angle to the axis of P_0 while the axis of P_2
is perpendicular to the axis of P_1. With such an arrangement the beam of radiation incident on
P_1 is divided by it into two coherent components with mutually orthogonal polarizations: The
component with an electric wave vector parallel to the lines of the polarizer is reflected from
P_1 while the component with a vector perpendicular to the lines passes through P_1. The radiation passing through P_1 after reflection from the corner reflector O_1 again enters P_1. Since
the axis of P_1 makes an angle of 45° with the edge of O_1, however, after reflection from O_1 the
plane of polarization of this beam is now turned through a 90° angle relative to the filaments
of P_1. As a result the beam is reflected from P_1 in the direction of O_1 and only after a second
reflection from O_1 (the plane of polarization of the beam is again turned 90°) it passes through
P_1 in the direction toward P_2, and after reflection from the latter (orthogonal polarization) it
is directed toward P. Thus, a path difference develops between the components of the beam
which initially fell on P_1, caused by the difference in the geometrical paths followed by each
of the divided beams. The important thing here is that the beam which passes through P_1 "cir-
culates" twice within the P_1-O_1 configuration. The component initially reflected from P_1 passes

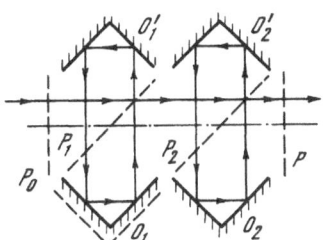

Fig. 8. Diagram of a twin-wave polarization
interferometer operating "on transmission."

through P_2 and will circulate twice within the P_2-O_2 system, after which it also travels toward P. By mounting the reflectors O_1 and O_2 on a movable platform, accomplishing their simultaneous plane-parallel displacement in the direction A-A' shown in Fig. 7, we obtain the possibility of regulating and varying the path difference of the interfering beams, with a displacement of the platform by a value a leading to a change of $6a$ in the path difference.

As in the interferometers examined in Section 2, in this interferometer a phase difference between the interfering beams leads to a change in the polarization of the beam emerging from the interferometer and is converted into amplitude variations in the signal using P. Thus, this system provides for obtaining path differences six times larger than the linear displacements of the mirror reflectors.

When it is necessary to obtain even larger path differences the polarizer P_1 is reoriented so that its axis becomes parallel to the edge of O_1. Then the path difference will be determined by the length of the path traveled by the beam during the circulation within P_2-O_2. By the choice of the value of a one can provide "overlapping" of the path differences obtained with the two different positions of P_1. During rotation of the polarizer P we obtain, as in [11], a variable signal proportional to the variable component of the interferogram.

The second interferometer variant, shown in Fig. 8, gives an example of a twin-beam interferometer which operates "on transmission" with amplitude division of the beam into coherent components. In a number of cases such a system allows one to obtain a more compact construction of the spectral instrument. In contrast to the preceding variant, in this construction four 90° corner reflectors are used. The axes of polarizers P_0 and P_1 are oriented parallel and at a 45° angle to the reflector edge. The axis of polarizer P_2 is set orthogonal to the axis of P_1. The ray path in the interferometer with reflections from P_1, P_2, and the corner reflectors is shown in Fig. 8. The analyzer P is again used to analyze the polarization of the emerging radiation. Changes in the phase difference of the beams are accomplished by variation in the distance between opposite corner reflectors.

The examination of the proposed polarization devices which was conducted allows one to conclude that they can prove effective in measurements of the spectra and polarization of sources in the submillimeter range.

In conclusion the author wishes to thank A. E. Salomonovich for interest in the work and assistance.

LITERATURE CITED

1. A. E. Salomonovich, Usp. Fiz. Nauk, 99, 427 (1969).
2. A. G. Kislyakov, Usp. Fiz. Nauk, 101, 607, (1970).
3. L. Mertz, Transformations in Optics, Wiley (1965).
4. P. B. Fellgett, J. Phys., 19, 187, 237 (1958).
5. High-Resolution Infrared Spectroscopy [Russian translation], Collection of articles edited by G. N. Zhizhin, Mir, Moscow (1972).
6. A. L. Fymat and K. D. Abhyankar, Appl. Opt., 9, 1075 (1970).
7. N. A. Irisova, Vest. Akad. Nauk SSSR, No. 10, 63 (1968).
8. A. G. Zhukov and V. I. Smirnov, Zh. Prikl. Spektr., 3, 410 (1960).
9. N. A. Irisova, Pis'ma na Zh. Éksperim. i Teor. Fiz., 2, 323 (1965).
10. A. I. Demeshina, V. A. Zayats, V. I. Lapshin, and V. N. Murzin, Zh. Prikl. Spektr., 13, 346 (1970).
11. D. H. Martin and E. Puplett, Infrared Phys., 10, 105 (1970).
12. M. Hass and M. O'Hara, Appl. Opt., 4, 1027 (1965).
13. J. B. Young, H. A. Graham, and E. W. Peterson, Appl. Opt., 4, 1023 (1965).

14. M. Czerný and H. Röder, Ergebn. Exact. Naturwiss., 17, 70 (1938).
15. H. A. Gebbie, J. E. Gibbs, J. E. Chamberlain, G. W. Chantry, F. D. Findlay, N. W. B. Stone, and A. J. Wright, Infrared Phys., 6, 195 (1966).
16. Yu. I. Kolesov, V. N. Listvin, and A. Ya. Smirnov, Preprint No. 118, Institute of Radio Engineering and Electronics, Academy of Sciences of the USSR, Moscow (1972).
17. L. A. Vainshtein, in: High-Power Electronics [in Russian], Izd-vo Akad. Nauk SSSR, Moscow (1963), No. 2, p. 26.
18. Z. S. Agranovich, V. A. Marchenko, and V. P. Shestopalov, Zh. Tekh. Fiz., 32, 381 (1962).
19. O. A. Tret'yakov and V. P. Shestopalov, Izv. Vuz. Radiofiz., 6, 352 (1963).
20. Long-Wave Infrared Spectroscopy [Russian translation], Collection of articles edited by V. N. Murzin, Mir, Moscow (1966).
21. R. C. Milward, Infrared Phys., 9, 53 (1969).
22. E. V. Ukhanov and O. K. Filippov, Opt.-Mekhan. Prom., No. 8, 69 (1971).
23. V. N. Listvin, Opt. i Spektr., 31, 151 (1971).
24. V. I. Lapshin and A. E. Salomonovich, Kratkie Soobshch. Fiz., Fiz. Inst. Akad. Nauk SSSR, No. 5, 51 (1971).
25. D. G. Vickers, E. I. Robson, and J. E. Beckman, Appl. Opt., 10, 682 (1971).
26. D. H. Martin, Infrared Detection Technique for Space Research, V. Mano and J. Ring, eds., Reprint, No. 12-854, (1971).
27. A. L. Fymat, Appl. Opt., 10, 2499, 2711 (1971); 11, 119 (1972).
28. W. Shercliff, Polarized Light [Russian translation], Mir, Moscow (1965).

ADJUSTED DEFORMATIONS OF MIRROR SYSTEMS OF
FULLY STEERABLE RADIO TELESCOPES

P. D. Kalachev, A. N. Kozlov, V. B. Tarasov,
and V. N. Titov

INTRODUCTION

The ever more complex problems of radio astronomy and distant, including space, radio communication require a further increase in the efficiency of radio telescopes, i.e., an increase in their resolving power and sensitivity.

An analysis of the various factors causing a reduction in the efficiency of fully steerable parabolic antennas of large radio telescopes shows that the main reason for the reduction in efficiency is the deviation of the real geometry of the mirror system from the ideal geometry.

In radio astronomy, where one can to a certain extent choose the time of the observations depending on the weather conditions, the main effect on the mirror system is the dead weight. To avoid the effect of nonuniform heating by solar radiation the observations can be conducted at night or in cloudy weather.

In space radio communication it becomes necessary to operate under any weather conditions. Therefore here, in addition to the dead weight, the mirror system is subject to the effect of wind loads at a substantial wind speed (on the order of $V = 20$ m/sec) and of the thermal radiation of the sun. Despite this, however, the dead weight loads are also the main external effects on the mirror system here.

The fact is that dead weight loads are in principle not removable at the earth's surface. The effect of nonuniform heating can be considerably decreased by a special coating for the constructions of the mirror system or fully removed by placing the radio telescope within a radio-transparent dome, in which case wind effects are also completely eliminated.

Thus, the main effect leading to distortion of the geometry of the mirror system is the effect of the dead weight.

In the general case of the direction of the antenna in space the mirror system is simultaneously loaded symmetrically and skew-symmetrically by the dead weight in accordance with the components into which the weight load can be resolved: P_1 and P_2 (Fig. 1).

With the optimum choice of the structual system of the supporting framework of the mirror the symmetrical loads lead to symmetrical deformations of the mirror, preserving it as a solid of revolution. In this case the curvature of the generatrix, generally speaking, can vary [1]. In application to parabolic mirrors the deformations for which the deformed mirror

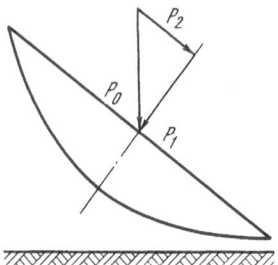

Fig. 1. Diagram of weight loading on a mirror.

remains a paraboloid of rotation and only the focal distance changes were called homologous deformations by S. von Hörner [2].

In connection with the recent development of methods of optimization of the radio engineering characteristics of parabolic fully steerable antennas, along with the problems solved by electrodynamics one of the important problems becomes the realization of the methods cited. One means of realizing the optimum characteristics of parabolic antennas is the use of structural systems which provide for adjusted deformations of the mirror system.

The concept of adjusted deformations of a mirror system and the means for their realization were introduced in [3, 4]. There the simplest case is examined — the prevention of longitudinal defocusing arising from the effect of the dead weight when the antenna is rotated about the elevation axis. A means of compensating for transverse displacements of the secondary (reradiating) mirror through the use of the moment of a counterweight, i.e., a means of achieving adjusted deformations of the mirror system in the presence of skew-symmetrical dead weight loads, is examined in [5]. On the basis of the methods of achieving adjusted deformation examined in [3-5] a more general determination of these deformations can now be formulated.

Deformations from the dead weight will be called adjusted deformations if the reduction in the antenna gain caused by them for any working elevation angle is compensated for by a change in the position of one or several functional elements of the mirror system caused by the effect of gravitational forces.

The general case of adjusted deformations of a mirror system is examined in the present report and a method is presented for the calculation of the values of the adjusted deformations when the values of the absolute deformations of the elements of the mirror system are known.

First of all we note that adjusted deformations not only ensure the absence of defocusing of the mirror system during rotation of the antenna about the elevation axis but also assure the retention of the focusing properties of the system achieved during the alignment of the reflecting surface of the mirror. The realization of adjusted deformations is the condition of minimum losses in the efficiency of parabolic antennas caused by deviation of the actual geometry of the mirror system from the designed geometry. Homologous deformations, which, generally speaking, presuppose the defocusing of the mirror system if automatic control of the position of the small mirror is not used, can be considered as a particular case of adjusted deformations.

In the development of new structural systems for the parabolic fully steerable antennas of reflecting radio telescopes it is necessary to make an estimate of the reduction of antenna efficiency caused by deviations of the actual geometry of the mirror system from the designed

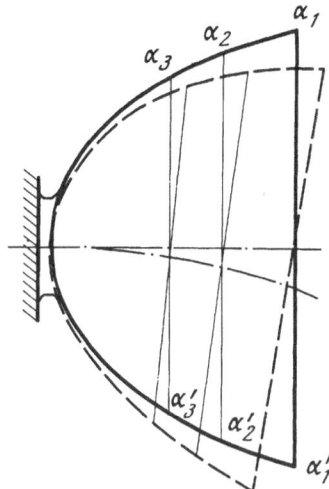

Fig. 2. Diagram of skew-symmetrical deformations of a mirror from the dead weight.

geometry. The deformation of the designed shape of the main mirror makes the main contribution to this loss. First of all, therefore, it is necessary to estimate the errors of the reflecting surface of the main mirror, which are comprised of technological errors δ_t in fabrication
and errors δ_d caused by deformations.

Current measuring techniques and fabrication technology permit one to attain very small
values of δ_t. Thus, for example, the 22-m mirror of the RT-22 of the Crimean Astrophysical
Observatory [6] was built with an error $\delta_t = 0.2$ mm (mean square); the 100-m mirror of the
Bonn University radio telescope was built with an accuracy $\delta_t \simeq 1$ mm (mean square). Prospects are seen for the further increase in measurement accuracy and consequently in the accuracy of alignment of the surface based on the use of laser techniques. Therefore the main
errors of a mirror surface are the errors due to deformations.

As already mentioned above, the dead weight simultaneously loads the mirror with
symmetrical and skew-symmetrical components (in the general case of antenna direction).
With the suspensions presently known the skew-symmetrical loading of a mirror by the dead
weight inevitably results in the fact that the mirror ceases to be a solid of revolution [7].

The deformed state of the mirror owing to the effect of the dead weight when the antenna
is directed toward the horizon is shown in Fig. 2. The deformed mirror is shown by the
dashed line. It is seen from Fig. 2 that the geometrical axis of the mirror has ceased to be
such but has been transformed into a curved line like the (camber) neutral axis of a beam subject to the effect of transverse bending.

The curvature of the axis can be decreased considerably by the choice of the optimum
structural system, although the known syspension systems, including multisupport systems,
do not fully eliminate the curving of the axis.

Skew-symmetrical loading of the mirror leads to the fact that the upper edge of the mirror in bending downward increases the curvature of the upper diametral cross section of the
mirror, while the lower edge, also bending downward, decreases the curvature of the lower
part of the mirror cross section. If one assumes that the deformed mirror has remained a
solid of revolution close to the undeformed mirror in shape, then its apex, at which the curvature

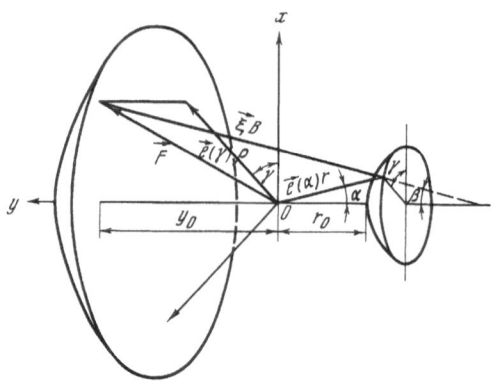

Fig. 3. Diagram of an axially symmetrical
two-mirror system.

is maximal, has been displaced upward. In this case a part of the mirror deformations can be
represented as an overall rotation of the mirror without distortion of its shape. This part is
quite considerable, on the order of 90-95% of the total deformation, so that only less than 1/10
of the absolute value of the deformations distorts the shape of the mirror.

1. METHODS OF PHASE COMPENSATION IN TWO-MIRROR

AXIALLY SYMMETRICAL ANTENNAS*

A schematic illustration of an axially symmetrical two-mirror system is shown in Fig. 3.
The following terms are adopted: ρ, γ, polar coordinates of a point in the aperture plane XOZ;
F, radius vector of surface of main mirror; r, α, polar coordinates of profile of secondary
mirror; y_0, distance from exciter to apex of main mirror; r_0, distance from exciter to
apex of secondary mirror; B, distance between mirrors along a reflected ray; β, angle between OY axis and direction of vector ξ; $\xi = j \cos\beta + e(\gamma) \sin \beta$, unit vector of a ray reflected
from secondary mirror; $e(\alpha) = i \sin \alpha - j \cos \alpha$, unit vector of a ray emerging from exciter;
$e(\gamma) = i \cos \gamma + k \sin \gamma$, unit-vector circular function in the aperture.

The following deformations, caused by skew-symmetrical loading of the mirror system
by dead weight forces, will lead to nonsymmetrical distortions of the wave front in the XOZ
plane:

1) a transverse displacement d_x of the apex of the main mirror;

2) rotation of the main mirror through an angle Φ;

3) a transverse displacement b_x of the apex of the secondary mirror;

4) rotation of the secondary mirror through an angle φ;

5) a transverse displacement a_x of the exciter;

6) rotation of the mirror system as a whole through an angle θ.

---

* In [8] this problem is considered in a geometrical sense.

Then, according to [9], the deviation of the wave front from a plane front at the point of the aperture with coordinates ρ, γ will equal

$$m\,(\rho,\ \gamma)=\mathbf{i}\,(\xi+\mathbf{j})\,d_x+(\xi+\mathbf{j})\,([\mathbf{Fk}]-iy_0)\,\Phi+\mathbf{i}\,(\mathbf{e}\,(\alpha)-\xi)\,b_x+$$
$$+(\mathbf{e}\,(\alpha)-\xi)\,(ir_0+r\,|\mathbf{ek}|)\,\varphi-(\mathbf{e}\,(\alpha)\,\mathbf{i})\,a_x+\rho\theta\cos\gamma. \tag{1}$$

It follows from Fig. 3 that $\mathbf{F}=\mathbf{e}\,(\alpha)\mathbf{r}+\boldsymbol{\xi}\,\mathbf{B}$. Expanding the scalar triple product $\boldsymbol{\xi}\,[\mathbf{FK}]$, we obtain

$$\xi\,[\mathbf{Fk}]=r\xi\,[\mathbf{ek}]. \tag{2}$$

Moreover,

$$\mathbf{j}\,[\mathbf{Fk}]=\mathbf{F}\,[\mathbf{kj}]=-(\mathbf{Fi})=-\rho\cos\gamma. \tag{3}$$

After substituting (2) and (3) into (1) and grouping terms, we find

$$m\,(\rho,\ \gamma)=(\xi\mathbf{i})\,(d_x-y_0\Phi-b_x-r_0\varphi)+(\mathbf{e}\,(\alpha)\,\mathbf{i})\,(b_x+r_0\varphi-a_x)+$$
$$+(\xi\,[\mathbf{ek}])\,r\,(\Phi-\varphi)+\rho\cos\gamma\,(\theta-\Phi). \tag{4}$$

In order not to decrease the antenna gain it is necessary to satisfy the equality $m(\rho, \gamma) = 0$ for all points of the aperture. This is possible only when the following conditions are satisfied:

$$\begin{aligned}
d_x-y_0\Phi-b_x-r_0\varphi&=0,\\
b_x-r_0\varphi-a_x&=0,\\
\Phi-\varphi&=0,\\
\theta-\Phi&=0,
\end{aligned} \tag{5}$$

which represent a system of equations for the phase compensation of skew-symmetrical deformations in two-mirror antennas. The rotation of the mirror system as a whole through an angle Φ which occurs in this case can be allowed for in the antenna aiming system.

If the antenna is constructed in such a way that the conditions (5) are satisfied for all working elevation angles θ, i.e., the law of variation of any deformation $P(\theta)$ will have the form

$$P\,(\theta)=P_m f\,(\theta), \tag{6}$$

where P_m is the amplitude and $f\,(\theta)$ is the law of variation with elevation angle which is the same for all the deformations considered, then one can speak of the adjusted nature of the deformations of such an antenna.

The condition $m(\rho, \gamma) = 0$ can be satisfied in the case when certain deformations equal zero. For example, in work at very short wavelengths with mirrors of small size, when because of the sufficient rigidity of the main mirror one can assume that $d_x = 0$, $\Phi = 0$, and $a_x = 0$, the construction of the secondary mirror mounting becomes the principal deforming unit and the system of equations (5) is reduced to the condition

$$r_0\varphi+b_x=0.$$

Then, equating (4) to zero and converting to the analytical form of notation, we obtain a differential equation for the profile of the secondary mirror. The solution of this equation is a sphere of radius $R = r_0 = b_x/\varphi$. The corresponding profile of the main mirror is easily calculated by the method of wave fronts and differs little from a parabolic form. If the rigidity of the suspension system of the secondary mirror is such that $b_x/\varphi = R = \text{const}$ with variation in the elevation angle, then there is no change in the gain and neither does a displacement of a ray occur in such an antenna.

Moving on to an estimate of the symmetrical deformations of a two-mirror antenna, in accordance with [9] one can write

$$m\,(\rho,\ \gamma) = j\,(\xi+j)\,d_y + j\,(e\,(\alpha)-\xi)\,b_y - (e\,(\alpha)\,j)\,a_j + j\,(\xi+j)\sum_n A_n \rho^{2n} + C = 0, \tag{7}$$

where d_y, b_y, and a_y are the longitudinal displacements of the apices of the main and secondary mirrors and the exciter, respectively, and $\sum_n A_n \rho^{2n}$ is the law of variation of the profile of the main mirror during symmetrical deformations.

The constant $c = 2\,(b_y - d_y) - a_y$ is determined from Eq. (7) with $\rho = 0$. Equation (7) is the condition of phase compensation of the deformations of a two-mirror antenna. The symmetrical deformations will be adjusted when the condition (6) is satisified.

Equation (7) can be satisified in two ways:

1) with the profiles of the main and secondary mirrors assigned and with deformations d_y, b_y, and a_y an appropriate law of variation of the profile of the main mirror is selected:

$$\sum_n A_n \rho^{2n} = -\frac{(b_y - a_y)\,(1 - \cos\alpha) + (b_y - d_y)\,(1 - \cos\beta)}{1 + \cos\beta}; \tag{8}$$

2) when the law of deformation of the profile of the main mirror is assigned Eq. (7) leads to a differential equation, from the solution of which the profiles of the secondary or main mirrors of the system can be determined. For example, when $b_y = a_y$, i.e., the system exciter – secondary mirror is displaced jointly, Eq. (8) is converted to the form

$$\sum_n A_n \rho^{2n} = -(b_y - d_y)\tan\frac{\beta}{2}.$$

From the law of reflection at the main mirror $\tan\frac{\beta}{2} = -\left(\frac{dy}{d\rho}\right)^2$. Then

$$\frac{dy}{d\rho} = \sqrt{\frac{\sum_n A_n \rho^{2n}}{b_y - d_y}}, \qquad y = \int \sqrt{\frac{\sum_n A_n \rho^{2n}}{b_y - d_y}}\,d\rho. \tag{9}$$

In particular, with the parabolic law of deformations $A_1 \rho^2$ which was investigated by S. von Hörner the solution of Eq. (9) will be

$$y = \frac{1}{2}\sqrt{\frac{A_1}{b_y - d_y}}\,\rho^2.$$

This is the equation of a parabolic mirror with a focal distance

$$f = \frac{1}{2}\sqrt{\frac{b_y - d_y}{A_1}}.$$

2. CONSTRUCTION OF APPROXIMATING SURFACE

The ways which were considered for the phase compensation of deformations in two-mirror antennas are realized if the main mirror in changing its position in space and its curvature remains a figure of revolution in all cases. As already mentioned, the existing means of suspension of the mirror do not preserve it as an axially symmetrical figure. Therefore, in the calculation of phase compensation instead of the real deformed main mirror one operates

with the concept of an axially symmetrical approximating surface, assigned by a set of coefficients which determine its shape and position in space. For brevity we will call them the approximation parameters.

The normalized gain $\varkappa$ in the presence of distortions of the wave front in an optimized
antenna with a uniform amplitude distribution in the aperture is determined by the equation

$$\varkappa = \frac{1}{\pi^2} \left| \int_s e^{-jkm} ds \right|^2,$$

where $k = 2\pi/\lambda$ and the integration is carried out over the area of the aperture. The radius
of the aperture is taken as unity. We use a series expansion of the exponential function and
limit ourselves to the first three terms. Then after several transformations we obtain

$$\varkappa \approx 1 - \frac{1}{\pi} \int_s (km)^2 \, ds + \frac{1}{\pi^2} \left[\int_s km \, ds \right]^2. \tag{10}$$

The parameters of the approximation are determined from the system of equations

$$\frac{\partial \varkappa}{\partial p_i} = 0,$$

where p_i is the i-th parameter of the approximation.

By differentiating (10) we obtain

$$\frac{1}{\pi} \int_s m \, ds \int_s \frac{\partial m}{\partial p_i} \, ds - \int_s m \frac{\partial m}{\partial p_i} \, ds = 0. \tag{11}$$

Let us represent the value m in the form

$$m = m_d + m_a = m_d + m_a^+ + m_a^-, \tag{12}$$

where m_d is the deviation of the wave front from a plane front in the deformed mirror, m_a is
the deviation of the wave front from a plane front in the approximating mirror, and m_a^+ and
m_a^- are the symmetrical and skew-symmetrical components of m_a, respectively.

Substituting (12) into (11), after several transformations with allowance for the parity
of the functions inside the integrals we find

$$\int_s m_d \frac{\partial m_a^-}{\partial p_i} \, ds = - \int_s \frac{\partial m_a^-}{\partial p_i} m_a^- ds, \tag{13}$$

$$\frac{1}{\pi} \int_s m_d \, ds \int_s \frac{\partial m_a^+}{\partial p_i} \, ds - \int_s m_d \frac{\partial m_a^+}{\partial p_i} \, ds = \int_s m_a^+ \frac{\partial m_a^+}{\partial p_i} \, ds - \frac{1}{\pi} \int_s m_a^+ ds \int_s \frac{\partial m_a^+}{\partial p_i} \, ds. \tag{14}$$

Equations (13) and (14) represent a system of linear equations from which the skew-symmetrical and symmetrical approximation parameters are determined, respectively.

Let us determine the values of m_d and m_a. As shown in [9],

$$m_d = (\mathbf{j} + \boldsymbol{\xi}) \, \Delta f_d = [\mathbf{j} \, (1 + \cos \beta) + \mathbf{e} \, (\gamma) \sin \beta] \, \Delta f_d,$$

where Δf_d is the deformation vector, determined by calculation or measured.

A typical case of the distribution of the components of the deformation vector at the point
with coordinates ρ, γ is shown in Fig. 4. We can write the vector Δf_d in the form $\Delta f_d =
-\mathbf{j} \Delta W + \mathbf{e}(\gamma) \Delta U + \mathbf{g}(\gamma) \Delta V$, where $\mathbf{g}(\gamma) = \frac{d\mathbf{e}(\gamma)}{d\gamma}$. Then

$$m_d = -\Delta W \, (1 + \cos \beta) + \Delta U \sin \beta. \tag{15}$$

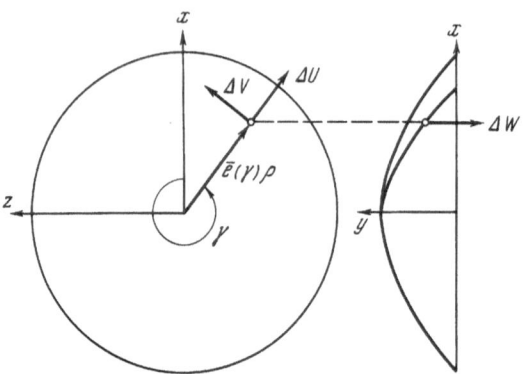

Fig. 4. Typical case of the distribution of components
of the deformation vector.

The value of m_a, determined through the approximation parameters d_x, d_y, Φ, and A_n in accordance with [9], is equal to

$$m_a = (j + \xi) \left\{ i d_x + j d_y + j \sum_n A_n \dot{\rho}^{2n} + [|Fk| - iy_0] \Phi \right\}. \tag{16}$$

The components m_a^+ and m_a^- calculated from (16) are, respectively, equal to

$$m_a^+ = (1 + \cos \beta) \left(d_y + \sum_n A_n \dot{\rho}^{2n} \right), \tag{17}$$

$$m_a^- = \cos \gamma \left\{ d_x \sin \beta + \Phi \left[(y - y_0) \sin \beta - \rho (1 + \cos \beta) \right] \right\}. \tag{18}$$

Solving the system of equations (13) and (14), after substituting into it Eqs. (15), (17), and (18) we obtain the values of the approximation parameters which provide for the construction of an approximating surface differing minimally from the deformed surface in the radio engineering sense.

The geometrical deviation Δn between the deformed and approximating surfaces, measured along the normal, can be calculated for each point from the equation

$$\Delta n = \frac{m^*}{2 \cos^2 \dfrac{\beta}{2}}, \tag{19}$$

where m^* is the result of the substitution into (12) of the values of m_a^+ and m_a^- determined from (17) and (18) with the approximation parameters obtained from Eqs. (13) and (14).

The change in $\varkappa$ produced by the deviations Δn can be calculated from Eq. (10).

3. RESULTS OF NUMERICAL CALCULATIONS

A calculation of the approximation parameters by the method presented was carried out for the main mirror of the RT-70 radio telescope with a diameter of 70 m. The deformations ΔW and ΔU were assigned for 133 control points of the mirror surface distributed in seven radial belts and 18 sectors (the step $\Delta\gamma = 10°$). In accordance with the initial data the maximum deviation of the deformed surface from the designed surface was 36 mm. Because of the finite number of control points the integration in Eqs. (13) and (14) was replaced by summation.

The calculated values of the approximation parameters are: $d_x = 128$ mm, $d_y = -1.5$ mm, $\Phi = 10'.3$, $A_1 = 10$ mm (which corresponds to a change of -14.4 mm in the focal distance). The rms deviation of the approximating surface from the deformed surface, determined from the mass of values Δn calculated from (19) for all the control points, is 0.86 mm. The corresponding reduction in the gain when operating at a wavelength of 3 cm, calculated from (10), is 4.5%.

It is interesting to note that a geometrical approximation based on the minimum in the deviation of the approximating surface from the deformed surface gives an rms deviation Δn equal to 0.8 mm. However, the reduction in the gain at the same wavelength which corresponds to this case is 5.5%.

CONCLUSION

The conditions (5)-(7) for the phase compensation of antenna deformations through the displacement of its functional elements from the effect of gravitational forces are derived from an examination of the phase relations in the aperture of a two-mirror system. These conditions allow one to rationally distribute the deformations among the functional elements in the construction of the antenna system and to choose a deformation law (8) for the profile of the main mirror which provides for their phase compensation.

In the determination of the parameters of the axially symmetrical surface which approximates the real deformed surface of the mirror the minimization criterion must be the size of the reduction in antenna gain. The system of equations for the determination of the approximation parameters of the surface is linear and decomposes into two subsystems for the symmetrical and skew-symmetrical parameters.

LITERATURE CITED

1. P. D. Kalachev, Tr. FIAN, 38, 73 (1967).
2. S. von Hörner, Astron. J., 72, 35 (1967).
3. P. D. Kalachev, Preprint FIAN, No. 171 (1968).
4. P. D. Kalachev, Tr. FIAN, 62, 150 (1972).
5. P. D. Kalachev and I. A. Emel'yanov, Tr. FIAN, 62, 136 (1972).
6. V. N. Ivanov, I. G. Moiseev, and Yu. G. Monin, Izv. Krim. Astrofiz. Obs., 38, 141 (1967).
7. P. D. Kalachev, Tr. FIAN, 38, 60 (1967).
8. P. D. Kalachev and M. V. Konyukov, Preprint FIAN, No. 149 (1971).
9. A. N. Kozlov, V. B. Tarasov, and V. N. Titov, Transactions of the Sixth Conference of Young Specialists of the A. S. Popov ITORÉS [in Russian] (1972), p. 163.

LIMITING DIMENSIONS OF A FULLY STEERABLE PARABOLIC
MIRROR FOR A RADIO TELESCOPE

P. D. Kalachev

For the solution of the ever more complex problems of radio astronomy it is necessary to create radio telescopes having the greatest possible resolving power and sensitivity. In addition to the corresponding demands on the radio receiving apparatus and the allowance for the wavelength range used, the above-mentioned demands on radio telescopes are connected with an increase in their dimensions. The question of the maximum possible size for the fully steerable parabolic mirror of a radio telescope subject to external effects under the conditions at the earth's surface arises in this connection.

The principal external effects to which the antenna system of a radio telescope* is subject are gravitational forces, wind, snow, icing, and the heating temperature. Inertial forces can be ignored since they are insignificant for radio telescopes.

There exist two limitations on the dimensions of a fully steerable parabolic mirror of a radio telescope: the limitation on the permissible deformations and the limitation imposed by the requirements of stability of the supporting construction. The first limitation depends on the admissible losses in the effective area of the mirror due to distortions in the paraboloidal form by elastic deformations. For a radio telescope intended for observations in the range of centimeter or even decimeter wavelengths the deformation limitation is stricter than the stability limitation. Since we are interested precisely in the maximum possible size (the diameter for a round aperture) of a fully steerable mirror we will analyze the problem of the maximum possible mirror diameter in view of the second limitation, i.e., based on the conditions of stability of the supporting structure.

In order not to complicate the problem we will consider only one external effect, gravitational forces, firstly because they are the primary effect (under the conditions of operation) [1], and secondly because all the other external disturbances (wind, snow, temperature) can in principle be eliminated, as by placing the radio telescope under a covering [2]. Thus, we will examine the question of the maximum possible mirror diameter on the basis of an analysis of the stresses in the supporting elements of the structure of the mirror and its suspension produced by the effect of the dead weight.

The formulation of the problem is complicated by the fact that the mirror of the radio telescope is movable: It must rotate about the two mutually perpendicular axes of the sup-

---

* Henceforth we will call it the mirror system, since in recent times a parabolic antenna, apart from the antenna-exciter proper such as a horn exciter, has most often consisted of a parabolic mirror and a small reradiating mirror located near the focus of the first mirror.

porting-rotating mechanism. To be specific we will examine the antenna system of a radio telescope with an altazimuth arrangement of the axes.

The rotation of the mirror system about the horizontal (elevation) axis introduces the main difficulty, since in this case the rotation produces a change in the direction of action of the gravitational forces. Stability of the mirror and its suspension must be assured for any position of it in space. However, it is sufficient to assure stability for two positions of the mirror: with its geometrical axis directed toward the zenith and toward the horizon. All other possible mirror positions can be reduced to the first two by resolving the weight forces into the corresponding directions.

First of all it must be noted that in solving the problem of the stability of the mirror the question is not only the stability of the mirror itself or even not so much its stability but rather, as will be shown below, the stability of the mirror suspension, since the latter is loaded not only by the weight of the mirror but also by the weight of the incomparably heavier supporting elements themselves. Consequently, we will examine the mirror system, including the mirror and its suspension.

Considering the various possible positions of the mirror system and wishing to simplify the problem, von Hörner [3] analyzed a three-dimensional system symmetrical relative to a horizontal axis of rotation and consisting of a rod octahedron. The stresses in the elements of this system from the dead weight are almost independent of its position relative to the horizontal axis. With identical cross sections for all the elements of the octahedron its maximum size (the distance between the outer supporting points), dependent on the dead weight loads, was obtained as:

a) $d_{max} \simeq 620$ m for a steel structure with an allowable steel stress $\sigma = 1400$ kg/cm^2 (ordinary steel used in building construction) and a specific weight $\gamma = 7.80 \cdot 10^{-3}$ kg/cm;

b) $d_{max} \simeq 1170$ m for a structure made of aluminum alloy with an allowable stress $\sigma = 910$ kg/cm^2 and a specific weight $\gamma = 2.7 \cdot 10^{-3}$ kg/cm^3.

Although the maximum size of the "mirror system" in the analysis cited above did prove to be close to the size obtained as a result of our analysis (see below), the replacement of a mirror system by the system of a rod octahedron gives a quite approximate solution of the formulated problem. The octahedron examined (as a mirror system) can serve only as the supporting part of a mirror suspension, and in this case not a full octahedron but a semiocta-hedron (the part of it below the horizontal plane containing the horizontal axis of rotation).

In a more exact solution of the formulated problem it is necessary to take into account the fact that the mirror system is not symmetrical relative to a plane parallel to the plane of the mirror aperture and containing the horizontal axis of rotation.

As will be shown below, the maximum possible size of a mirror loaded by the dead weight depends essentially on the mirror suspension system.

Two mirror suspension schemes are examined below: a cantilever system and a multi-support system with a radially symmetrical arrangement of the supports.

SCHEME WITH A CANTILEVER MIRROR SUSPENSION

A radial supporting element for the mirror framework in the form of a cantilever, fixed at the left end which represents the center of the mirror, is shown in Fig. 1. The length of the cantilever is equal to half the mirror diameter, 0.5D; the structural height h_0 at the root is equal to the depth of the mirror: $h_0 = (0.5D)^2/4$ or $h_0 \simeq 0.208D$ for F = 0.3D. With symmetrical loading of the mirror by the dead weight we take the elements shown by the dashed lines

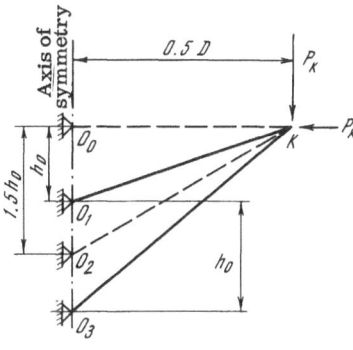

Fig. 1

O_0K and O_2K as the cantilever elements; with skew-symmetrical loading the cantilever elements are those shown by solid lines: O_1K and O_3K. This corresponds to the fact that with symmetrical loading we must allow for the work of the annular (chord) elements, while with skew-symmetrical loading they can be neglected.

Symmetrical Loading. In this case a radial element is loaded not only by its own weight but by the weight of all the structural elements and of the chord (annular) elements which do not take part in receiving the total loads (the work of the annular elements is imitated as it were by the element O_0K which has a horizontal rather than an inclined direction).

On the basis of preliminary studies we adopt a load coefficient, which takes into account the additional weight of the overlying elements, equal to 2.5. Assuming in the first approximation that all the total load is distributed uniformly along the element, we obtain the load at the end of the cantilever:

$$P_\kappa = 0.5\bar{K}_1(0.5D + \sqrt{0.25D^2 + 2.25h_0^2}\,F\gamma) \cong 0.550\bar{K}_1 D\gamma E.$$

The forces in the elements are $N_{0,\kappa} = P_\kappa/\sin\varphi = 0.55\,D\bar{K}_1 F\gamma/\sin\varphi_1$. $N_{0_0\kappa} = 0.55\,\bar{K}F\gamma/\tan\varphi$. The stresses in the elements are $\sigma_{0,\kappa} = N_{0,\kappa}/F$, $\sigma_{0_0\kappa} = N_{0_0\kappa}/F$. Taking the angle between the horizontal and inclined rods (Fig. 1) as $\varphi = 32°$ from structural considerations, we obtain

$$\sigma_{0,\kappa} = 1.03D\gamma\bar{K}_1, \qquad \sigma_{0_0\kappa} = 0.873D\gamma\bar{K}_1. \tag{1}$$

Skew-Symmetrical Loading. In this case, first, the direction of the load P_k is changed by 90°, and second, the unit load $\bar{P}_K$ increases by $\bar{K}_2$ times due to the weight of the neighboring radial elements which are located farther from the vertical plane of symmetry [4]. The value of the coefficient $\bar{K}_2$ depends on the number of supporting radial elements for the mirror frame; for large mirrors $\bar{K}_2 \simeq 1.99$.

With skew-symmetrical loading the scheme shown by the solid lines (see Fig. 1) operates.

The load at the end point K of the cantilever is

$$P_K = 0.5F\gamma\bar{K}_1\bar{K}_2(l_{0,\kappa} + l_{0_0\kappa}),$$

where $l_{0,\kappa} = 0.54D$, $l_{0_0\kappa} = 0.65D$, and $P_\kappa = 0.6F\gamma\bar{K}_1\bar{K}_2D$. The forces in the supporting elements are

$$N_{0,\kappa} = \frac{P_\kappa}{\cos\alpha\,(1 - \tan\alpha\cdot\cot\psi)}, \qquad N_{0_0\kappa} = \frac{P_\kappa}{\sin\psi\cdot\cot\alpha - \cos\psi}.$$

For $\alpha = 22°.5$, $\psi = 40°$ (see Fig. 1), $K_1 = 2.5$, and $K_2 = 1.99$ we obtain

$$\sigma_{0,\bar{R}} = \frac{N_{0,\bar{R}}}{F} = 5.98 D\gamma, \qquad \sigma_{0,R} = \frac{N_{0,R}}{F} = 3.89 D\gamma. \tag{2}$$

If the structure of the mirror system is built of steel then the maximum size of the main mirror proves to be equal to (from (2))

$$D_{max} = \frac{\sigma \, (cm)}{5.98\gamma} = 298 \text{ m.}$$

Taking different cross sections (for equalization of the stresses) $F_{0_1\bar{R}} = 1.5 F_0$ and $F_{0_3\bar{R}} = F_0$, we obtain

$$P_{\bar{R}} = 0.5 F_0 \gamma K_1 K_2 (l_{0,\bar{R}} 1.5 + l_{0,R}) = 3.64 D F_0 \gamma.$$

The stresses will be $\sigma_{0,\bar{R}} = 4.89 \, D\gamma$, $\sigma_{0,R} = 4.76 \, D\gamma$.

Adopting the same material (steel) in the construction of the mirror system, we find $D_{max} = 1440 \cdot 10^3/4.89 \cdot 7.85 \simeq 365$ m, i.e., a slight equalization of the stresses (an increase in the strength of the weak element) permitted a sharp increase in the size of the mirror.

Full equalization of the stresses, i.e., the optimum distribution of cross sections of the supporting elements of the structure, can lead to an increase in the maximum possible mirror diameter to about $D_{max} \simeq 460$ m.

SCHEME OF A MIRROR SYSTEM WITH A MULTISUPPORT

MIRROR SUSPENSION

Let us consider a mirror system in which the parabolic mirror has a multisupport suspension with a radially symmetrical arrangement of the supports, i.e., represents a three-dimensional circular system.

The equation of the parabola forming the mirror profile has the form

$$y = \frac{x^2}{4F}, \tag{3}$$

where F is the focal distance of the mirror, which to be definite we will take as equal to 0.3D from now on (D is the diameter of the mirror).

Force diagrams of the mirror system are presented in Figs. 2 and 3. The rods shown by dashed lines are located only in the back panels. The structural elements (intermediate chord elements, braces, the nodes for fastening the facing, and the facing itself which forms the reflecting surface of the mirror) are not shown in the diagrams. The total weight of the structural elements usually comprises ~15-35% of the total weight of the mirror. Also not shown in the diagram is the structure supporting the exciter system, since its weight does not exceed 5% of the weight of the mirror, and with a successful construction can be reduced to 1-2%.

The main geometrical proportions of this scheme are

$$\begin{aligned}
&a = 0.125D, &m = 0.03D, &\quad C = 0.005D, \\
&b = 0.375D, &t = 0.063D, &\quad l = 0.315D, \\
&h = 0.115D, &C_1 = 0.010D, &\quad l_0 = 0.22D, \\
&B = 0.5D.
\end{aligned}$$

The sizes of the load-bearing elements of the structure can be expressed analogously. For example, the sizes of the chord elements (l_{ii}) are determined from the equation

$$l_{ii} = 2\rho_i \sin \psi. \tag{4}$$

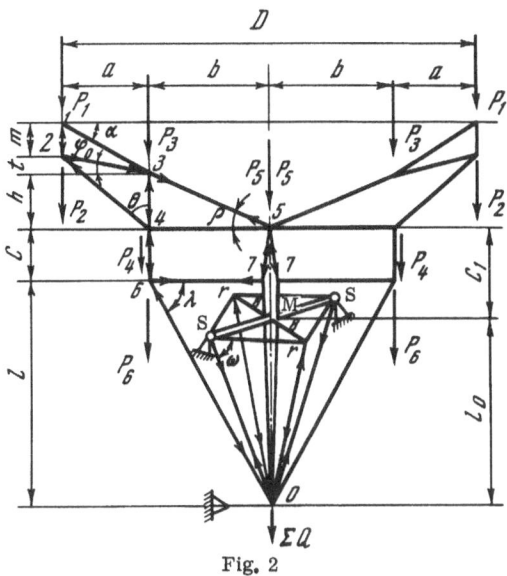

Fig. 2

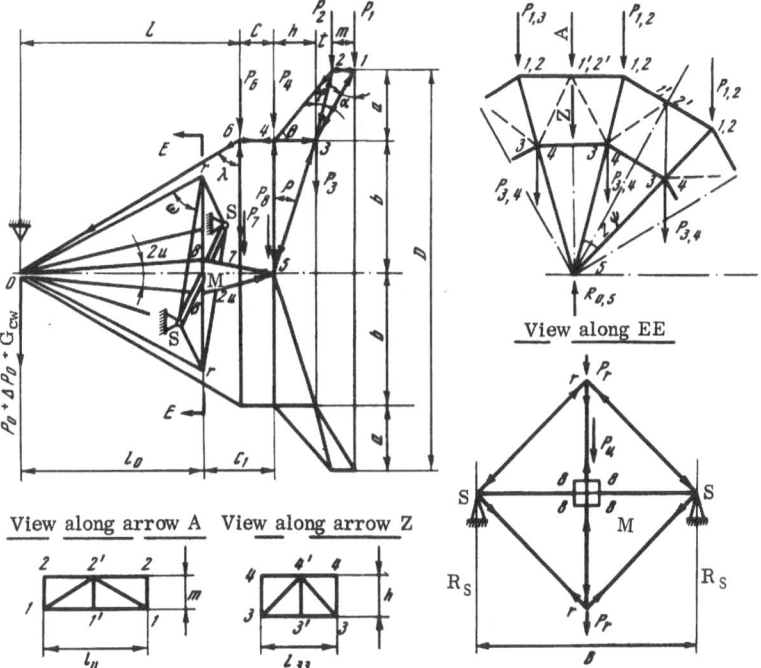

View along EE

View along arrow A   View along arrow Z

Fig. 3

Here $\psi = 360°/2n$ and n is the number of radial load-bearing elements of the mirror framework, which we take* as equal to 32. Consequently, $\psi = 5°37'30''$ and $l_{ii} = 0.196\,\rho_i$, where ρ_i is the distance of the chord elements from the center of the mirror.

It should be noted that for the maximum possible mirror diameter determined by the condition of stability it is necessary to assure the equal stability of all (or at least of the principal) supporting elements of the mirror system. Consequently, the cross sections of the load-bearing elements must be chosen in accordance with the forces acting in them. But since the forces in the supporting elements are proportional to the effective loads and the latter in turn are porportional to the cross sections of the elements, the problem can be solved only by the method of successive approximations.

In the first approximation we will start from the assumption that the areas of the cross sections of all the load-bearing elements are the same. With this assumption the nodal load (from the dead weight) is determined by the equation

$$P_i = 0.5 \bar{K} F \gamma \sum_1^n l_n, \tag{5}$$

where F is the cross-sectional area of an element, cm^2; l_n is the length of the n-th element, cm; γ is the specific weight of the construction material, kg/cm^3; n_1 is the number of elements (rods) converging at the given node. The coefficient K allows for the additional load (overload) of the given supporting element. This coefficient has different numerical values for different elements and for different positions of the mirror in space, i.e., with different loadings of the mirror system by the dead weight: symmetrical and skew-symmetrical.

When $\bar{K} = 1$ Eq. (5) gives the nodal load produced only by the dead weight of the supporting elements but not allowing for the weight of the structural elements not shown in the diagrams of Figs. 2 and 3 (the facing which forms the reflecting surface of the mirror; the intermediate elements serving for the fastening of the facing; the elements of the bracing system which assure the effective working of the load-bearing elements under compression forces).

Since the facing and the structural elements of the mirror framework serving for the fastening of the facing are located in the front (face) part of the mirror, the overloading of the corresponding (front) nodes of the mirror framework is accordingly greater than the overloading of the back nodes.

Considering what has been said above, we adopt a different coefficient $\bar{K}_1$ for different nodes: $\bar{K}_1 = 1.45$ for nodes 2, 1, and 3 (see Fig. 2) and $\bar{K}_1 = 1.25$ for the other nodes, extending this coefficient not only to the elements of the mirror proper but to all the elements of its suspension, i.e., to all the elements of the system (except for nodes 1, 2, and 3).

In addition to the coefficient $\bar{K}_1$, the coefficient $\bar{K}$ in Eq. (5) also takes into account the coefficients $\bar{K}_2$ and $\bar{K}_3$ of the overloads produced in the chords and radial elements during skew-symmetrical loading.

The magnitude of the coefficient $\bar{K}_2$ of the overloading in the chord elements depends on the number of radial supporting elements in the mirror framework [4]:

$$K_2 = \frac{1}{\cos\psi}\left[\cos\psi + \sum_2^i \cos(2i - 1)\right]. \tag{6}$$

---

* The number n is taken as minimal based on the assumption that the maximum diameter of the mirror will be less than 1000 m.

In our case with n = 32 the number of supporting radial elements i in one quadrant of the mirror framework is 0.25n = 8, and

$$K_2 = \frac{1}{\cos \psi}[\cos \psi + \cos 3\psi + \cos 5\psi + \ldots + \cos 15 \psi] = 8 \cos 8\psi \cdot \cos 4\psi \cdot \cos 2\psi \cdot \cos \psi.$$

Substituting the appropriate value of ψ, we obtain $\overline{K}_2$ = 5.122.

The magnitude of the coefficient $\overline{K}_3$ pertaining to the radial elements also depends on i and is determined from the equation [4]

$$K_3 = 2 \sin (2i - 1) \psi. \tag{7}$$

In our case i = 8 and $\overline{K}_3$ = 2 sin 15ψ ≈ 1.99.

The nodal load from the dead weight, found on the basis of (5), has the following values:

$$
\begin{aligned}
P_1 &= 0.383FD\gamma, & P_5 &= 0.442FD\gamma, \\
P_2 &= 0.849FD\gamma, & P_6 &= 0.590FD\gamma, \\
P_3 &= 0.949FD\gamma, & P_7 &= 0.216FD\gamma, \\
P_4 &= 0.956FD\gamma, & P_0 &= 0.284FD\gamma, \\
P &= 0.742FD\gamma, & K_1 &= 1.15.
\end{aligned}
$$

The load at the middle node M is determined from the expression

$$P_M = n \left(\sum_1^7 P_i + P_0 \right) + G_{se} + G_{cw}, \tag{8}$$

where

$$n \left(\sum_1^7 P_i + P_0 \right) = 32 (P_1 + P_2 + \ldots + P_7 + P_0) \approx 149.8FD\gamma,$$

$$G_{se} = F\gamma [4 (l_{58} + l_{08}) + 2 (0.5l_{0r}) + 2 (0.5l_{0s}) + 0.5l_{ss}]$$

is the weight of the suspension elements.

The size of the counterweight G_{cw} is determined from the conditions of balancing the system relative to the horizontal (elevation) axis of rotation:

$$n [P_1 (m + t + h + C_1) + P_2 (t + h + C_1) + P_3 (n + C_1) + C_1 (P_4 + P_5) +$$

$$+ (C_1 - C) (P_6 + P_7)] + \Delta P_5 l_1 = n P_0 l_0 + \Delta P_0 l_0 + l_0 G_{cw} \tag{9}$$

where ΔP_5 = 0.5Fγ (4l_{58}); ΔP = 0.5Fγ (4l_{08} + 2l_{0r} + 2l_{0S}). After the appropriate substitution and transformations we obtain from (9) G_{cw} = 102.44FDγ.

Substituting the corresponding values into (8), we obtain

$$P_M = 254.83FD\gamma.$$

The forces in the supporting elements of the system are determined from the equations of statics:

$$
\begin{aligned}
\Sigma X &= 0, & \Sigma Y &= 0, & \Sigma Z &= 0, \\
\Sigma M_x &= 0, & \Sigma M_r &= 0, & \Sigma M_z &= 0.
\end{aligned}
\tag{10}
$$

The forces in elements 58 and 08 of the suspension are determined by the equations:

TABLE 1

Designation of nodal loads P_i	Value of nodal load in kg × $DF_0\gamma$ for four approximations				Designation of nodal loads P_i	Value of nodal load in kg × $DF_0\gamma$ for four approximations			
	1st	2nd	3rd	4th		1st	2nd	3rd	4th
P_1	0.383	0.995	0.923	0.889	P_r	0.742	159.7	428.61	1104.00
P_2	0.849	3.636	4.422	5.147	ΔP_0	0.897	318.1	719.10	1435.00
P_3	0.949	6.247	10.320	17.740	ΔP_5	0.246	53.24	116.10	226.20
P_4	0.956	2.952	3.600	4.350	G_{se}	2.332	704.8	1638.10	4780.00
P_5	0.442	3.582	5.813	8.310	G_{cw}	102.44	210.2	60.10	—260
P_6	0.590	8.551	14.210	29.731	P_M	254.826	1949.1	3361.82	7880.00
P_7	0.216	2.377	4.751	8.631	ΔP_0	0.666	202.6	436.30	932.50
P_0	0.284	3.982	7.962	14.200	ΣQ	246.69	319.4	2434.84	5776.50

a) with symmetrical loading*

$$N_{58} = \frac{h\left(\sum_1^7 P_i - N_{46}\right) + \Delta P_5}{4 \cos U''}, \qquad N_{08} = 0, \tag{11}$$

where $\cos U'' = (C_1/l_{58}) \tan U'$;

b) with skew-symmetrical loading

$$N_{58} = \frac{n\left(\sum_1^7 P_i\right) + \Delta P_5}{4 \sin U''}, \qquad N_{08} = \frac{C\left[n\left(\sum_1^7 P_i\right) + \Delta P_5\right]}{4 l_0 \sin U}. \tag{12}$$

The forces in the main supporting elements $\overline{OS}$ with symmetrical loading are determined from the equation

$$N_{0s} = \frac{\Sigma Q}{2 \sin \omega}, \tag{13}$$

where

$$\Sigma Q = (N_{06} \sin \lambda + P_0) n + 2P_r + \Delta P_0' + G_{cw},$$
$$\Delta P_0' = 0.5 F \gamma (2l_{0r} + 2l_{08}).$$

The forces in elements $\overline{OS}$ with skew-symmetrical loading are equal to zero.

The forces N_{ii} in the chord elements with symmetrical loading are taken as equal to zero, which went into the margin of stability of the radial elements which are somewhat relieved by the work of the chord elements.

The forces in the chord elements with skew-symmetrical loading are found from the equation

$$N_{ii} = P_i \overline{K}_2, \tag{14}$$

where $\overline{K}_2$ is determined by Eq. (6).

The nodal loading P from the dead weight, calculated from Eq. (5) for the first, second, third, and fourth approximations corresponding to successive refinements of the cross-sectional areas of the supporting elements of the structure, is presented in Table 1. The stresses in

---

* There is no longitudinal reaction (along the geometrical axis of the mirror) at node O.

TABLE 2

Designation of element	Area of cross section F_0F_i, cm^2	Symmetrical loading	Skew-symmetrical loading	Mirror diameter D = 800 m	
		Fourth approximation		Sym.	Skew-sym.
		$\sigma/D\gamma$, kg/cm^2	$\sigma/D\gamma$, kg/cm^2	σ, kg/cm^2	σ, kg/cm^2
1—2	1	0.889	1.321	559	830
1—3	1.7	0	1.308	0	822
2—3	12	0.613	1.608	385	1010
2—4	7.6	1.498	1.610	939	1011
3—4	9	1.828	0.746	1148	469
3—5	32	0.215	1.776	135	115
4—5	5	1.314	0.064	826	40
4—6	18	1.674	0.986	1052	620
0—6	50	1.876	0.557	1179	350
6—7	40	1.806	2.018	1133	1268
5—8	800	0.144	2.013	91	1265
0—8	680	~ 0	2.018	0	1268
r—M	1800	~ 0	2.188	~ 0	1250
1—1	4	~ 0	1.132	~ 0	711
2—2	16	~ 0	1.648	~ 0	1036
3—3	46	~ 0	1.978	~ 0	1243
4—4	13.5	~ 0	1.651	~ 0	1039
6—6	80	~ 0	1.662	~ 0	1045
r—S	1700		2.100		1224
0—s	2000	2.183	~ 0	1373	~ 0
0—r	800	2.087	~ 0	1311	~ 0
S—S	7000			1371	

the supporting elements of the mirror system of a radio telescope from the effect of the dead weight, obtained in the fourth approximation, are presented in Table 2. The stresses were determined from the equation $\sigma = N_{in}/F_{in}$, where $N_{in} = P_i K_2$ for both cases of stress (symmetrical and skew-symmetrical). It is seen from the table that in the fourth approximation the stresses are about the same in all the supporting elements of the mirror system from the effect of the dead weight. Further equalization of the stresses in the supporting elements, as the corresponding calculations show, have little effect on the value of the maximum possible mirror size.

The maximum possible size of the mirror is determined by the stability of the weakest link. This link in the given case is element OS, i.e., the main supporting element of the suspension (see Fig. 2) for the case of symmetrical loading. For this element the stress is $\sigma = 2.183D\gamma$, from which $D \simeq 0.458(\sigma_{OS}/\gamma)$. For $\sigma = 1400$ kg/cm^2 (ordinary structural steel) and $\gamma = 7.85 \cdot 10^{-3}$ kg/cm^3 we have $D_{max} = 0.458(1400$ cm$/7.85 \cdot 10^{-3}) \simeq 835$ m. Keeping in mind that we are determining D_{max} as an estimate, one can assume with a certain margin of safety that $D_{max} = 800$ m.

The stresses in the supporting elements calculated for a mirror diameter $D_{max} = 800$ m are presented in Table 2.

When the mirror system is built of aluminum alloy with an allowable stress $\sigma = 910$ kg/cm^2 and a specific weight $\gamma = 2.7 \cdot 10^{-3}$ kg/cm^3 the maximum mirror diameter will be $D_{max} = 1540$ m, i.e., almost two times greater than with steel construction.

Thus, if one assures the stability of the mirror system only from the loads of the dead weight of the construction then the maximum possible size of a fully steerable parabolic mirror on a multisupport suspension proves to equal $D_{max} = 800$ and 1500 m for steel and aluminum constructions, respectively.

As for the other external effects besides gravitation, as indicated earlier the radio telescope can be placed inside a covering (for example, within a protective dome which is transparent in the range of radio waves) and the influence of such effects as wind, snow, and heating will be removed. Even without a covering, however, the construction scheme which we examined for the mirror system of a radio telescope with a main parabolic mirror 800 m in diameter, calculated only for the effect of gravitational forces, can also take a wind load.

The allowable velocity is determined from the following considerations. Suppose the maximum stress in a supporting element of the structure which develops during the combined effect of forces of gravitation and a wind load on the mirror system must not exceed the allowable stress by more than 10%. Such an increase in the stress is justified by the fact that even a wind with a low velocity, for example 10 m/sec, occurs only 10% of the time in a year, i.e., the effect of a wind load is short-lived. Then for the determination of the allowable wind load we obtain the equation

$$q_r = 0,1 q_G, \tag{15}$$

where $q_G = \Sigma G_{ms}/S_m$ is the weight of 1 m² of the mirror aperture, ΣG_{ms} is the total weight of the mirror system, and S_m is the area of the mirror aperture. The wind load is $q_v = C \, \rho (V^2/2)$, where $\rho = 0.125$ (at sea level), $C_x = 1.6$ is the aerodynamic resistance coefficient of the mirror system, and V is the wind velocity.

Substituting the appropriate values into (15), we obtain

$$V = \sqrt{q_G}. \tag{16}$$

The weight of 1 m² of the mirror aperture is a function of the normalized value F_0 of the cross-sectional area through which the cross sections of all the supporting elements are expressed (see Table 2). And since the value of F_0 can be limited only from below, i.e., F_0 cannot be less than a certain value determined by structural considerations (the allowable stability of long rods working in compression, for example),* it follows from (16) that if, for example, $q_G = 50, 100,$ and 150 kg/m² then the allowable wind velocity will equal V = 7, 10, and 12 m/sec, respectively.

In principle one can design and put into operation a heavier mirror system with a mirror diameter of 800 m whose structure is able to take wind loads in the presence of a hurricane force wind equal to V = 40 m/sec. In this case the allowable stress can be taken as increased: $\sigma = 2000$ kg/cm², and the weight of 1 m² will be $q_G = V^2 (\sigma/\sigma_x) = 1600(1400/2000) = 1120$ kg/m², i.e., in order for a mirror system to be insensitive to the effect of a hurricane wind its weight proves to be an order of magnitude greater than the weight of a mirror system calculated only for the dead weight load. Evidently the construction of a special "transparent" domed covering would be advisable.

LITERATURE CITED

1. P. D. Kalachev, Tr. FIAN, 28, 51 (1965).
2. H. G. Weiss, IEEE Spectrum, 54, 5 (1965).
3. S. von Hörner, Astron. J., 72, 35 (1967).
4. P. D. Kalachev, Tr. FIAN, 38, 60 (1967).

EXPERIMENTAL STUDY OF STRUCTURAL SYSTEMS OF AERODYNAMIC COMPENSATORS IN APPLICATION TO PARABOLIC ANTENNAS

V. E. D'yachkov, S. L. Myslivets, and V. P. Nazarov

In the practice of designing antenna-rotating systems (ARS) having a parabolic mirror great attention is paid to problems of decreasing the aerodynamic moments on the aiming and pinning mechanisms either by the introduction of certain systems of arrangement of the antenna aiming axes or by installing special devices, called aerodynamic compensators, on the mirror.

Aerodynamic compensation of the moment characteristics of ARS makes it possible to increase: 1) the allowable working wind velocity for the aiming mechanisms; 2) the working life of the aiming mechanisms; 3) the dimensions of the reflecting surface of the mirror when it is mounted on existing supporting-rotating mechanisms; 4) the accuracy in aiming the antenna assembly through a decrease in the mirror deformations; 5) the "hardiness" of the ARS during the maximum winds corresponding to the region of its installation.

In addition to shifting the aiming axes and the creation of a perforated surface along the edge of the reflector an effective means for decreasing the moment characteristics of an ARS with a parabolic mirror is a shield mounted on the reflector along its perimeter without a clearance and compensators mounted on the counterweight arms of the tilting part of the antenna. This method of aerodynamic compensation was confirmed experimentally in the aerodynamic laboratory of the Department of Hydro- and Aerodynamics of the Leningrad Pedagogical Institute. A diagram of the arrangement and dimensions of the shield (flap) and compensators and the results of the aerodynamic trimming are presented in Fig. 1 and in Table 1. The Oxyz coordinate system is a flow coordinate system whose Ox axis coincides with the velocity vector V of the oncoming flow; β is the elevation angle and α is the angle of yaw of the model relative to the direction of flow.

The installation of a flap consisting of two sections mounted at an angle δ to the tangent at the edge of the reflector (variant 1) led to a decrease of 1.53 times in the positive moment relative to the azimuthal axis and of 1.26 times in the negative moment compared with a model without a flap (Fig. 2). A drawback of this variant is the large size of the flap along the generatrix ($\bar{b} = 0.300$). For example, for an ARS of average size with a mirror diameter $D_r = 25$ m the width of the flap would be 7.5 m. It does not seem possible to put such a recommendation into practice. Therefore in the further trimming of the ARS model with a no-clearance flap we went to the simultaneous use of compensators on the counterweight arms by gradually varying their dimensions. This made it possible to decrease $\bar{b}$ to 0.175 with the same effect of a decrease in the maximum aerodynamic moments (Fig. 3).

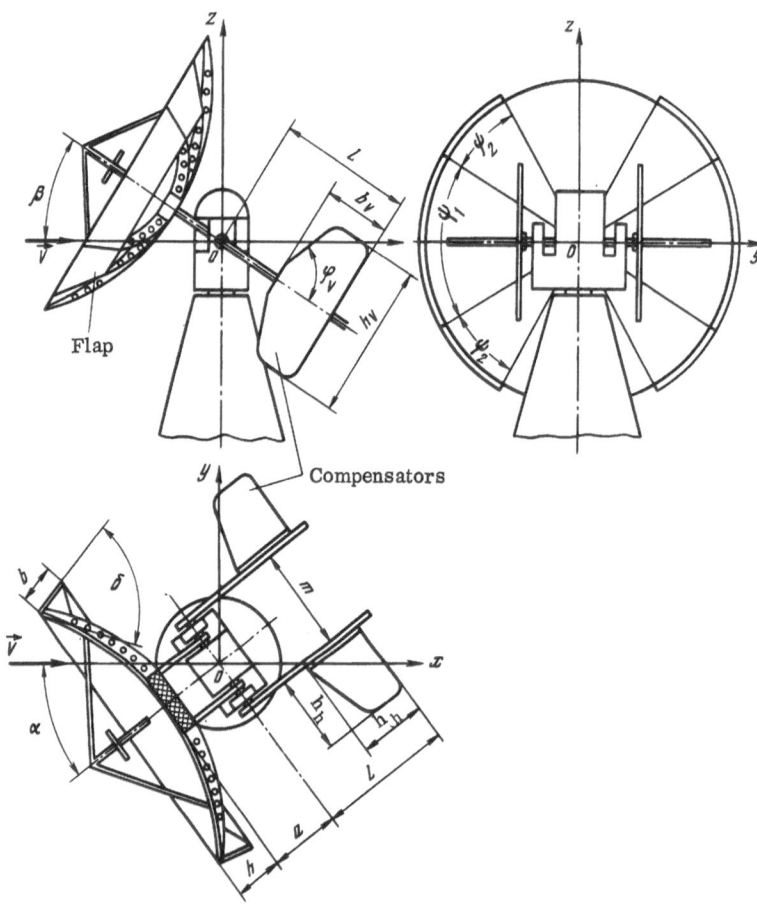

Fig. 1. Diagram of mounting of no-clearance flap and compensators on ARS model.

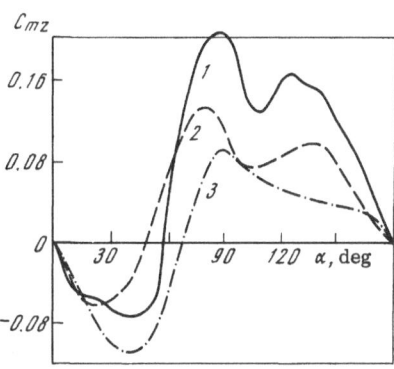

Fig. 2. Comparative results of aerodynamic trimming in azimuth for an ARS model using a no-clearance flap and compensators. 1) Without flap or compensators (variant 0); 2) with no-clearance flap (variant 1); 3) with no-clearance flap and compensators (variant 2). $\beta = 0$, $\bar{h} = 0.185$, $\bar{a} = 0.210$.

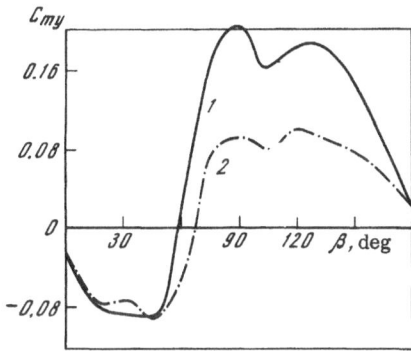

Fig. 3. Comparative results of aerodynamic trimming in elevation for an ARS model using a no-clearance flap and compensators. 1) Without flap or compensators (variant 0); 2) with no-clearance flap and compensators (variant 2). $\alpha = 0$, $\bar{h} = 0.185$, $\bar{a} = 0.210$.

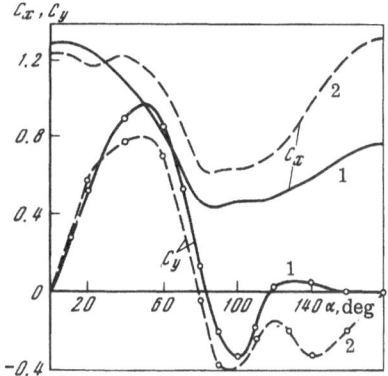

Fig. 4. Coefficients of frontal resistance C_x and lateral force C_y for variants 0 and 2. 1) Without flap or compensators (variant 0); 2) with no-clearance flap and compensators (variant 2). $\beta = 0$, $\bar{h} = 0.185$.

TABLE 1. Dimensions of No-Clearance Flap and Compensators Divided

Variant	Principal changes compared with previous variant	Dimensions of flap		
		b	ψ_1 (δ), deg	ψ_2 (δ), deg
0	Flap and compensators absent	—	—	—
1	Flap installed	0.300	60 (40)	30 (20)
2	Dimensions of flap changed and compensators installed	0.175	360 (70)	—

As a comparative analysis of the aerodynamic force coefficients shows (Fig. 4), the installation of the flap and compensators in accordance with variant 2 led to the practical conservation of the coefficients of lateral force C_y and even to their decrease in the region of maximum values. The coefficient of frontal resistance C_x increased in almost the entire interval of angles α. An especially appreciable increase in C_x by about 1.7 times took place for $\alpha = 140\text{-}180°$. For variant 2 the coefficient C_x reached a maximum not with frontal blowing on the reflector but with the stream coming from its back side, $C_{x\,max} = 1.34$. Attention should also be drawn to the increase by about 1.4 times in the coefficient C_x when $\beta = 0$ and $\alpha = 90°$. This antenna position is identical in the nature of the flow to the pinned position when the antenna installation is subjected to the effect of storm winds.

Among the other drawbacks of the method of aerodynamic compensation under consideration one can include:

1) the large overall sizes of the flap and especially of the compensators on the counterweight arms of the tilting part;

2) the considerable distance of the flap and compensators from the ARS aiming axes, which leads to an increase in the dynamic loads on the aiming mechanisms when they operate in tracking or transfer modes from the corresponding accelerations and to worsening of the dynamic quality of the ARS;

3) an increase in deformations of the reflecting surface and a consequent decrease in the effective area of the antenna due to the increase in the wind and weight loads on the edge of the mirror.

The effect of equalization of the pressure along the generatrix of the reflecting surface of the antenna owing to the creation of a clearance between the flap and this surface was first noted during aerodynamic trimming of an antenna of the Rakovina type. Despite the apparent difference in the shape of the reflecting surfaces (a combination of cut-outs from a paraboloid and a hyperboloid of rotation for the Rakovina type model and cut-outs from a paraboloid of rotation in our case) the authors of the present report suggested the mounting of a flap with a clearance on the back side of the parabolic reflector. This was an effective means of decreasing the positive part of the moment graph with the peaks of its negative part and of the diagram of frontal resistance being kept unchanged.

For an explanation of the nature of the flow of an air stream over a reflector with a flap of this type installed (Fig. 5) one can use experimental curves of the pressure distribution over the surface of a reflector without a flap. A typical curve in this respect is that (Fig. 6a) which occurs with $\beta = 0$ and $\alpha = 90°$ or when $\beta = 90°$ and $\alpha = 0$ or $180°$, when the aerodynamic moment relative to the azimuthal 0z and elevation 0y aiming axes is maximal or close to it (depending on the depth of the reflector). Assuming that the distribution of excess pressure

by Diameter of Reflector, and Results of Aerodynamic Trimming

Dimensions of compensators								Aerodynamic coefficients of moments			
		horizontal			vertical						
l	m	b_h	h_h	φh, deg	b_v	h_v	φv, deg	$(C_{my})_{max}$	$(C_{my})_{min}$	$(C_{mz})_{max}$	$(C_{mz})_{min}$
—	—	—	—	—	—	—	—	0.204	—0.089	0.206	—0.077
—	—	—	—	—	—	—	—	—	—	0.135	—0.061
0.45	0.43	0.25	0.25	70	0.25	0.45	70	0.097	—0.090	0.093	—0.708

$\bar{P}_e^{\pi}$ over the concave surface of the reflector is constant, the reduction in the moment can be explained by the following factors:

1) the operation of the flap in conjunction with the windward part of the reflector on the principle of a convergent nozzle;

2) the operation of the flap in conjunction with the leeward part of the reflector on the principle of a divergent nozzle;

3) the countermoment from the resultant force of the flow pressure directly on the conical surface of the flap.

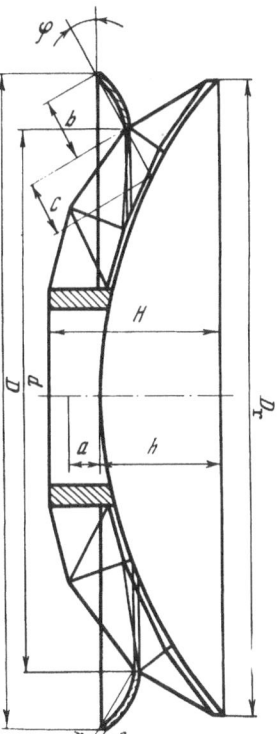

Fig. 5. Diagram of mounting of a flap with a clearance on the back side of a parabolic reflector model.

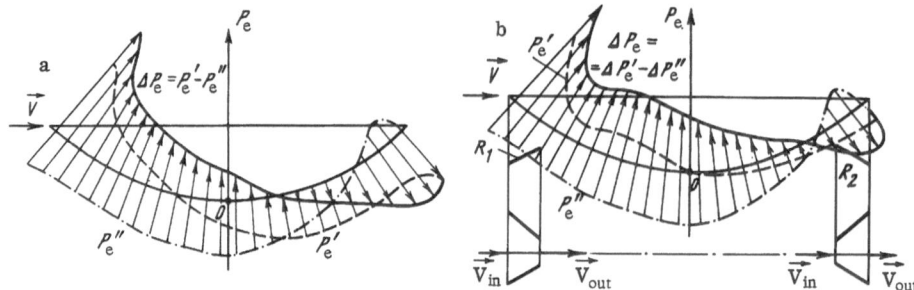

Fig. 6. Nature of distribution of excess over convex (P_e') and concave (P_e'') surfaces of reflector. a) For reflector model without flap; b) for reflector model with a flap mounted on the back side with a clearance. For convergent nozzle: $V_{out} > V_{in}$ and $P_{out} < P_{in}$. For divergent nozzle: $V_{out} < V_{in}$ and $P_{out} > P_{in}$.

When a stream flows in a convergent nozzle its velocity along the nozzle axis increases while the pressure decreases. In a divergent nozzle it is the opposite. This analogy of the operation of a flap set with a clearance explains the decrease in the compression pressure in the region of the leading edge and in the rarefaction pressure near the trailing edge, which ultimately leads to the equalization of the pressure $\overline{P}_e'$ on the back side of the reflector and to the shifting of the point of zero pressure ($\Delta P_e' = 0$) toward the trailing edge (Fig. 6, b). The quantitative estimate of the effect of equalization of the pressure on the convex surface of the reflector with such a method of compensation of the moment characteristics of an ARS is naturally subject to experimental confirmation in an aerodynamic laboratory.

The dimensions of the flaps for their different variants and the results of the aerodynamic trimming are presented in Table 2. The flaps for variants 1-12 had flat meridional cross sections while for variants 13-16 the cross sections were of parabolic shape.

TABLE 2. Dimensions of Flaps with a Clearance and Results of Aerodynamic Trimming

Variants	Dimensions of parabolic reflector and its supporting structure				Dimensions of flap						Aerodynamic coefficients of moments	
	D_T, mm	h, mm	a, mm	H, mm	D, mm	d, mm	b, mm	φ, deg.	c, mm	δ, mm	$(C_{mz})_{min}$	$(C_{mz})_{max}$
0					—	—	—	—	—	—	−0.066	0.138
1					537	449	49		0		−0.057	0.112
2					537	480	33		16		−0.058	0.109
3					551	449	57		0		−0.052	0.109
4					551	464	49	30	8	0	−0.054	0.103
5					565	449	65		0		−0.055	0.108
6					565	464	57		8		−0.065	0.100
7					565	480	49		16		−0.069	0.101
8	537	112	26.5	148	543			20			−0.058	0.104
9					539			25			−0.063	0.096
10					537	445	52	30	23	0	−0.062	0.078
11					530			35			−0.070	0.095
12					524			40			−0.054	0.089
13						374	100	35	38		−0.078	0.064
14					537	402	84	36	35	2	−0.076	0.068
15						430	68	37	32		−0.070	0.077
16						458	52	38	29		−0.055	0.087

The variable parameters which determine the dimensions of the flap and its position relative to the reflector are:

1) outer diameter D and inner diameter d of flap;

2) angle of taper φ of flap;

3) width c of clearance between flap and back side of reflector in the direction of the flap generatrix;

4) depth of camber of the f-generatrix.

The width of the flap is determined through the variable parameters according to the equation

$$b = \frac{D - d}{2 \cos \varphi}.$$

The most effective means of decreasing the maximum positive moment while preserving the negative moment at the same level, which is generally characteristic with this method of aerodynamic compensation, is to decrease the angle of taper φ of the flap with the parameters d, b, and c held constant (Fig. 7). The optimum angle of taper for the test model of a reflector

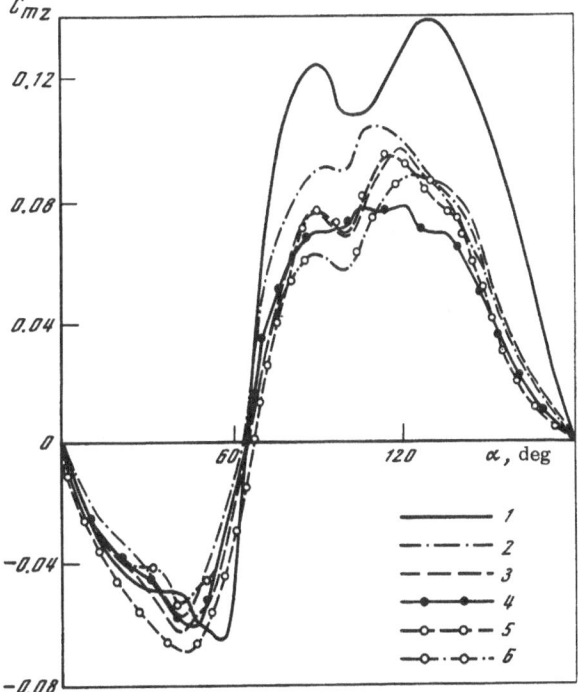

Fig. 7. Effect of angle of taper φ of a flap with a clearance on the moment coefficient C_{mz} ($\beta = 0$, $\bar{h} = 0.208$, $\bar{a} = 0.049$). 1) Reflector without a flap; reflector with a flap; 2) variant 8 ($\varphi = 20°$); 3) 9 ($\varphi = 25°$); 4) 10 ($\varphi = 30°$); 5) 11 ($\varphi = 35°$); 6) 12 ($\varphi = 40°$).

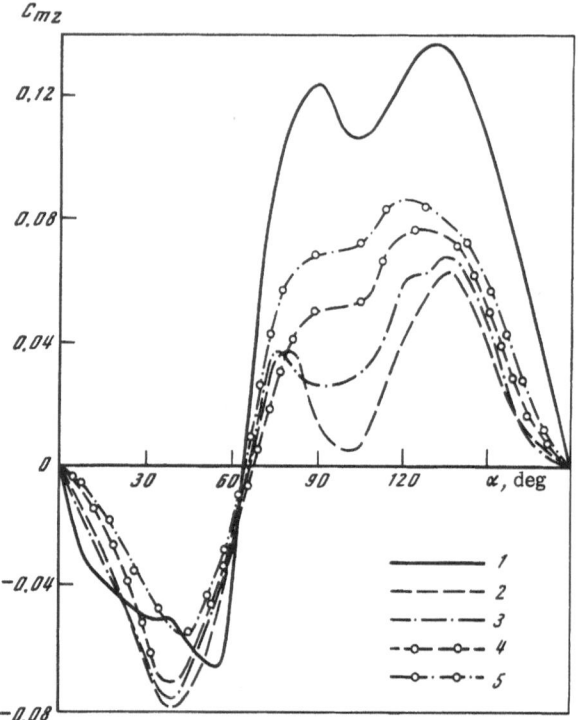

Fig. 8. Aerodynamic moment coefficients C_{mz} for variants of flaps with a parabolic generatrix ($\beta = 0$, $\bar{h} = 0.208$, $\bar{a} = 0.049$). 1) Reflector without a flap; reflector with a flap; 2) variant 13; 3) 14; 4) 15; 5) 16.

with a relative depth $\bar{h} = 0.208$ should be taken as $\varphi = 30°$. For shallower reflectors the angle will evidently be somewhat less.

As an analysis of the data presented in Table 2 shows, with an increase in the relative clearance $\bar{c} = c/D_r$ from 0 to 0.03 and with a simultaneous decrease in the relative width $\bar{b} = b/D_r$ of the flap but with the equality $\bar{b} + \bar{c} = \text{const}$ maintained, the coefficient $(C_{mz})_{max}$ decreases by 7% while $(C_{mz})_{max}$ increases by 25% in absolute value (variants 5-7). With an increase in the relative width $\bar{b}$ of the flap, going beyond the limits of the reflector aperture, and with $\bar{c}$ held constant, $(C_{mz})_{max}$ decreases, and more considerably for variants with a large clearance. However, in connection with the smallness of this effect and with the striving to decrease the dimensions of the flap, variation in this parameter was omitted in subsequent studies.

From the results of tests of the flaps in variants 13-16, for which the generatrix is made in the form of a parabola for the purpose of the smoother entry of the stream into the space between the leading edge of the reflector and the flap and of producing the required rigidity of the flap, it follows that with the outer diameter of the flap held constant and equal to the mirror diameter an increase in the width b and consequently in the clearance c has the result that the peak of the positive moment becomes less than the peak of the negative moment (Fig. 8). By interpolation one can select a variant of the flap for which these peaks are equal. Also characteristic is the considerable decrease (by four to eight times) in the positive moment when

$\beta = 0$ and $\alpha = 90°$ (or when $\beta = 90°$ with different α) for variants 13 and 14, which makes it possible to erect ARS in regions with stronger storm winds when flaps of such configuration and size are used. But the main disadvantage of the flaps of these variants is their ever greater overall size ($\bar{b} = 0.16$-0.19).

Of all the 16 variants of flaps mounted on a reflector with $h = 0.208$ which were tested the most acceptable was variant 10 ($\bar{D} = 1$, $\bar{b} = 0.097$, $\bar{c} = 0.043$, $\varphi = 30°$), for which the reduction in the moment characteristic in the positive region was 43% with the peak of the negative moment being almost preserved. Thus, in comparison with a no-clearance flap the same aerodynamic effect is obtained with a flap of smaller dimensions mounted with a clearance on the back side of the reflector. In this case there is no need to mount aerodynamic compensators on the counterweights of the tilting part.

The use of a flap with a clearance is an effective means of aerodynamic balancing both for two-axis ARS and for ARS with more than two operating axes.

In conclusion let us dwell on the nature of the change in the force coefficients for a reflector with the flap in variant 10 and compare them with the loads on a reflector without a flap (Fig. 9):

1) the coefficient of frontal resistance C_x with the direction of flow into the bowl of the reflector ($\alpha = 0$, $\beta = 0°$) and onto the reflector from the side ($\alpha = 90$, $\beta = 0°$) remained almost unchanged;

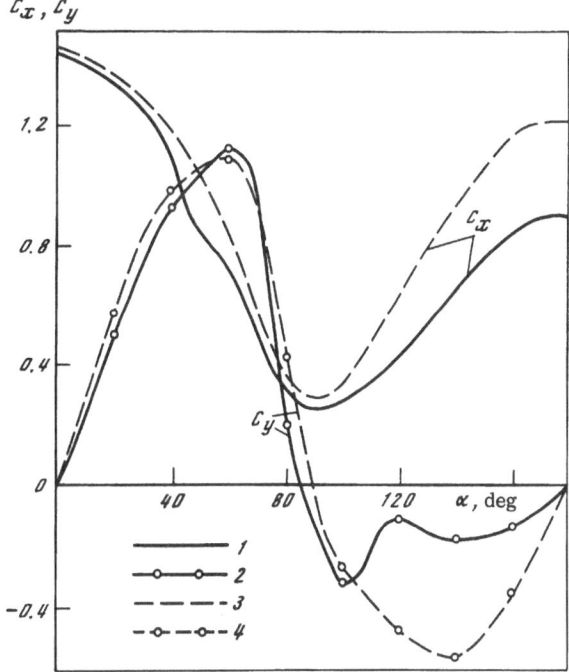

Fig. 9. Coefficients of frontal resistance C_x and lateral force C_y for variants 0 and 10 ($\beta = 0$, $\bar{h} = 0.208$). 1, 2) Without a flap (variant 0); 3, 4) with a flap installed with a clearance on the back side of the reflector (variant 10).

2) with the direction of flow from the back side of the reflector ($\beta = 0$, $\alpha = 180°$) the coefficient C_x increased to 1.207 in place of $C_x = 0.906$ for a reflector without a flap;

3) the maximum positive coefficient of lateral force did not change, and while its negative value did increase in absolute value by 1.8 times it remained considerably less than the positive value;

4) with $\beta = 0$ and $\alpha = 90°$ the value of C_y, or the coefficient of lift C_z which is the same thing (at $\beta = 90°$), changed by 1.6 times on the side of a decrease.

From what has been presented one can draw the following conclusions:

1. Of all the methods presently known for the aerodynamic compensation of the moment characteristics of ARS having a parabolic mirror the installation of a flap with a clearance on the back side of the reflector provides the greatest effectiveness in performing the assigned task.

2. The presence of a clearance alters the pattern of flow of a wind stream over the reflector, which results in more uniformity of the pressure curve over the surface of the reflector and consequently a decrease in the deformations of this surface and an increase in the sensitivity (effective area) of the ARS. The latter fact plays an especially important role since deformations of the mirror from the wind load cannot be compensated for, whereas weight deformations can be almost fully compensated for by known methods of homology.

3. The method of aerodynamic compensation using a flap with a clearance allows one to obtain the same reduction (by ~43%) in the aerodynamic moments relative to the operating axes of the antenna as with the use of a no-clearance flap in conjunction with compensators on the counterweight arms of the tilting part but with smaller flap dimensions and the complete elimination of the compensators. This method also extends to the wind balancing of three-axis ARS.

4. As a result of the modernization of existing ARS through the installation of a flap with a clearance on the back side of the reflector one can:

a) increase the maximum working wind velocity to

$$V'_w = \sqrt{n}\, V_w = 1.32 V_w,$$

where $n = 1/(1 - k) = 1.75$ is the efficiency of the moment compensation and $k = 0.43$ is the reduction in the moment relative to the operating axes of the antenna established as a result of the aerodynamic trimming;

b) erect them in a region with greater storm winds ($V'_{max} = 1.32 V_{max}$);

c) increase their resolving power (aperture area) if on a supporting-rotating mechanism designed for the wind loads determined by the diameter D_r of a reflector without a flap one installs a reflector of larger diameter equipped with the proposed flap: $D'_r = \sqrt{n}\, D_r = 1.2 D_r$.

5. The preeminence of the method of aerodynamic compensation using the installation of a flap with a clearance does not exclude its use in conjunction with other methods, which can be the subject of further studies in this direction.

A PARABOLIC RADIO TELESCOPE ANTENNA WITH
A RADIALLY BALANCED MAIN MIRROR

P. D. Kalachev, V. P. Nazarov, I. A. Emel'yanov,
V. L. Shubeko, and V. B. Khavaev

As is known, the main problem which must be solved in the creation of fully steerable parabolic antennas of high resolving power in the short centimeter and millimeter wavelength ranges is the problem of the geometry of the mirror system, i.e., the problem of providing high focusing properties for the main parabolic mirror and preserving them when the antenna is rotated about the elevation axis of the supporting-rotating mechanism. In this case the refocusing of the mirror system should not occur when the antenna rotates about the horizontal axis of the supporting-rotating mechanism as a result of displacements of elements of the exciter system in the longitudinal and transverse directions.

At a certain time a break developed between the advances of electrodynamics in the study of problems of the dependence of the efficiency of parabolic antennas of reflecting radio telescopes on the sizes of the errors in the reflecting surface of the mirror and the lag in the development of problems of the optimum structures for parabolic antennas, without which it is impossible to implement the advances in electrodynamics.

The problems of electrodynamics mentioned have been worked out by many authors. The most complete modern working out of these problems is contained in the works of Shifrin [1]. An approximate (final) equation without allowance for the effect of the correlation radius, which somewhat exaggerates the effect of surface errors on the efficiency losses of parabolic antennas, is presented in Ruze's report [2]:

$$\Delta W = 1 - \exp\left[-(4\pi\bar{\varepsilon}/\lambda)^2\right],$$

where $\bar{\varepsilon}$ is the rms value of the deviations of the real surface from the calculated surface and λ is the wavelength. Here $\bar{\varepsilon}$ includes the technological error $\bar{\delta}_t$ in the fabrication of the reflecting surface of the mirror and the error $\bar{\delta}_d$ caused by elastic deformations of the mirror. The most probable value of $\bar{\varepsilon}$ can be represented by the equation

$$\bar{\varepsilon} = \sqrt{\bar{\delta}_t^2 + \bar{\delta}_d^2}.$$

The current measuring methods and fabrication technology permit one to attain very small values of $\bar{\delta}_t$. For example, a value of $\bar{\delta}_t$ equal to ~ 0.2 mm was attained in the fabrication of the 22-meter parabolic mirror of the RT-22 radio telescope of the Crimean Astrophysical Observatory (1965) [3]. The corresponding value of $\bar{\delta}_t$ for the 100-meter mirror of the radio telescope of Bonn University (1971) was equal to ~ 1 mm [4].

As for the errors produced by deformations, they can be an order of magnitude greater if the structural scheme of the mirror system does not provide for the elimination of skew-symmetrical deformations, i.e., the preservation of the main mirror as a paraboloid of rotation after its deformations from the effect of the dead weight.

The improvement of the focusing properties of the main mirror and their preservation during the rotation of the antenna about the horizontal axis are and continue to be the principal reserve for increasing the efficiency of this type of antenna.

The preservation of the focusing properties of the mirror during rotation in elevation is connected with the elastic properties of the mirror structure, i.e., with its rigidity. The problem is not fully solved by a simple increase in the rigidity of the main mirror since it would be necessary to make both the main mirror itself and its suspension (as well as the suspension of the secondary mirror and the exciter, i.e., the basic elements of the mirror system) absolutely rigid. But this is impossible because the physical constants of all known construction materials, i.e., the modulus of elasticity E and the specific weight γ, have finite values.

It was necessary to take a fresh approach to the problems of the construction of parabolic antennas. In particular, the new approach to the construction of parabolic antennas which was proposed and developed in the laboratory of radio astronomy of the P. N. Lebedev Institute of Physics of the Academy of Sciences of the USSR (IPAS) is based on the principle of adjusted deformations of elements of the mirror system [5].

The use of the principle of adjusted deformations, just like that of homologous deformations which can be considered as a particular case of adjusted deformations (when there is refocusing of the mirror system), was connected with the necessity of developing and using new structural schemes for parabolic antennas which would provide for the fullest realization of this principle.

It must be noted that not one of the presently known structural schemes for a parabolic mirror and its suspension, including multisupport suspensions having a radially symmetrical arrangement of the supports, provides for the full realization of adjusted deformations. This is explained by the fact that the nature of skew-symmetrical deformations of a parabolic mirror from dead weight loads (antenna directed toward the horizon), even in optimum schemes from the point of view of the distribution of supports, prevents one from preserving the deformed parabolic mirror with high accuracy as a paraboloid of rotation. The question concerns deviations on the order of millimeters (for a mirror size of about 100 m) between the deformed surface of the mirror and a paraboloid of rotation. The fact is that existing schemes are not able (in principle) to eliminate the small curvature of the geometrical axis of the mirror which arises during skew-symmetrical loading. When the antenna is directed toward the horizon the geometrical axis of the mirror is bent like the curved axis of a cantilever. In Fig. 1 the deformed state of a mirror subject to skew-symmetrical loading by the dead weight is shown by a thin line.

While this curvature of the geometrical axis of the mirror is insignificant for long-focus mirrors (mirrors of small depth), for short-focus mirrors (F/D < 0.35), which are being used more and more in such antennas, this curvature becomes noticeable. Finally, this pertains primarily to large antennas (70-100 m) intended for work in the short centimeter and millimeter wavelength ranges [4, 6].

Our latest studies of the structural schemes of mirror systems have led to the development of a fundamentally new original scheme with a radially balanced main mirror. The creation of this system resulted from the attempt to get rid of the above-mentioned apparently unremovable defect of previous systems — the curvature of the geometrical axis.

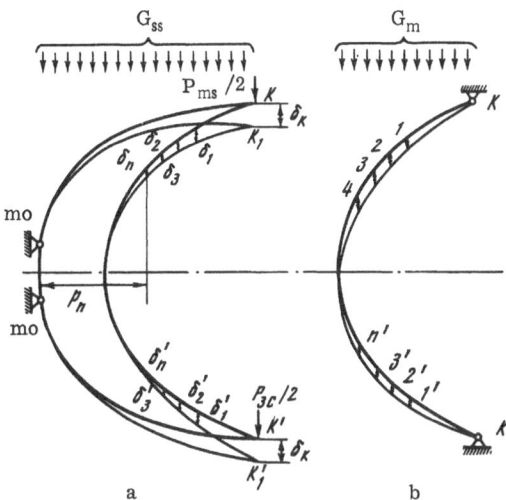

Fig. 1. Diagram of skew-symmetrical deformations of
the load-bearing supporting structure (a) and framework
of the mirror proper (b).

The new structural scheme for a mirror system, which we call the scheme with a radially balanced main (parabolic) mirror, in principle provides for the elimination of curvature of the geometrical axis. In addition, the new scheme allows one to provide any position (inclination) of the geometrical axis of the mirror (within certain limits, of course) during skew-symmetrical loading by the dead weight (antenna directed toward the horizon).

The new scheme for the suspension of a fully steerable parabolic (main) mirror also is a multisupport system with a radially symmetrical arrangement of the supports. However, the principal description of the mirror and the arrangement of the supporting elements of the suspension differ radically from all the presently known schemes, namely:

1) the principal fastening of the main mirror occurs along the outer boundary (the exact opposite to the scheme with a bracket suspension when the mirror is fastened to the central part of the supporting framework, the so-called central hub);

2) the supporting elements of the suspension (rods working both in tension and in compression) are arranged radially symmetrically in planes parallel to the plane of the mirror aperture;

3) the radial supporting elements work mainly during skew-symmetrical loading.

The scheme with a radially balanced main mirror contains two basic units: the supporting load-bearing structure and the framework of the mirror proper. In addition, the scheme has the system of radial supporting elements (braces) mentioned above. The overall dimensions of the load-bearing (supporting) structure exceed the dimensions of the mirror proper. The load-bearing structure is, so to speak, the mount for the mirror, similar to an optical telescope.

The suspension of the load-bearing structure itself on the supporting-rotating mechanism is accomplished by the bracket method, i.e., from its central part. Here the radially balanced scheme for the main mirror suspension does not impose any restrictions or special requirements on the dimensions of the supporting-rotating mechanism, which in this case can have either a vertical base (the tower type) or a horizontal base (such as a rotating platform on rollers or dollies).

The radial balancing of the main mirror is based on the principle of adjusted deformations of the load-bearing structure, the framework of the mirror proper, and the supporting radial elements.

Let us examine the qualitative nature of the skew-symmetrical deformations separately for the load-bearing structure loaded by the dead weight and by the weight of the mirror proper when the antenna is directed toward the horizon, supported along the boundary, and loaded by the dead weight. The load-bearing structure and the mirror framework proper are shown in Fig. 1. Their deformed states are shown by thin lines. Because of the symmetry of the structure and the skew-symmetry of the load the deflections of the corresponding points (nodes) of the upper and lower halves of the load-bearing structure are the same: $\delta_k = \delta_k^I$, $\delta_1 = \delta_1^I$, ..., $\delta_n = \delta_n^I$. On the other hand, the magnitude δ_n of the deflection increases with an increase in the distance of the node from the mountings of the load-bearing structure in proportion to the distance p_n to the power m:

$$\delta_n = k p_n^m,$$

where m > 1.

Furthermore, if one imagines that the framework of the mirror proper is fastened only along the outer boundary then from the effect of dead-weight skew-symmetrical loads its geometrical axis also ceases to be such after the deformations but is curved, in the same way as occurs in the load-bearing structure, in accordance with the symmetry of the structure and the skew-symmetry of the load. In this case the displacements will also increase in proportion to the distance p_n of the node from the point of fastening of the mirror to the power n:

$$\delta_n' = k_1 \cdot p_n^n, \ \ldots$$

If the mirror framework were rigidly fastened to the load-bearing structure at each node, denoted in Fig. 1 by the numbers 1, 2, 3, ..., n, then after deformation the framework would take on the shape of the deformed load-bearing structure, since the rigidity of the latter is many times greater than the rigidity of the mirror. Consequently, in this case the deformed mirror would have the form shown by the thin line, i.e., the mirror deformations would occur in the opposite direction from those which occur when the mirror is fastened along the outer boundary. It is not hard to show that if the corresponding nodes of the mirror framework and the load-bearing structure are not joined rigidly but through elastic supporting elements then by the appropriate choice of their elasticity (rigidity against longitudinal forces, i.e., radial forces*) one can provide for the desired shape of the deformed mirror, and in particular preserve the straightness and desired direction of its geometrical axis.

We note that with skew-symmetrical loading the general rotation of the entire mirror system will be observed owing to deformations of the load-bearing structure, but this does not affect the focusing properties of the mirror and only requires the introduction of a systematic correction to the readings of the instruments for reckoning the direction of the antenna in elevation.

The scheme of the arrangement of radial supporting elements in one of the planes parallel to the mirror aperture is shown in Fig. 2.

---

* The supporting elements, as mentioned above, are arranged in radial directions in planes parallel to the plane of the mirror aperture.

Fig. 2. Plan view of the scheme of a cross section in a plane
parallel to the aperture.

The rigidity of the radial supporting elements against longitudinal loads (along the element) must satisfy the condition

$$(\Delta l_n)_{\mathrm{r.e}} = \delta_k - \delta_{n'}.$$

The satisfying of this condition means that the parabolic mirror retains the shape of a paraboloid of rotation during skew-symmetrical loading. As for the focal distance, the latter must change when the antenna is rotated about the horizontal axis in accordance with the change in the distance between the main and reradiating mirrors, i.e., deformations of the mirror system must be satisfied [5].

On the basis of the properties of cyclic systems [7-9] — the preservation of the plane cross sections parallel to the mirror aperture and of the proportionality of the forces in the radial elements to the sine of the angle α between the direction of the radial element and the horizontal axis of symmetry of the main mirror (the axis of skew-symmetry of the loads) — we can assert that the forces in the radial supporting elements parallel to the horizontal plane of skew-symmetry ($\sin \alpha = 0$) are equal to zero and increase to a maximum for the radial element located in the vertical cross section. This means that a radial supporting element is loaded by the forces (from the weight of the mirror) in accordance with the position of the element (its inclination to the horizontal axis), i.e., the deformations of the load-bearing structure, the mirror framework, and the radial supporting elements in any other (slanting) cross section are adjusted just as in the vertical cross section under consideration.

Thus, in principle this radially balanced structural scheme assures the preservation of the shape of the mirror as a paraboloid of rotation in the presence of skew-symmetrical loading. However, the first trial static calculations show that the attainment of high accuracy in the preservation of the shape of a paraboloid of rotation requires the repeated checking of many variants with the selection of the appropriate rigidity for all the elements of the structure, which can be done only with the use of modern computer technology and consequently with the development of the appropriate programs.

Numerical studies of a structural scheme with a radially balanced parabolic mirror 128 m in diameter having a relative focal distance of 0.35D (F = 44.8 m) are presently being conducted in the radio astronomy laboratory (radio telescope sector) of the IPAS.

According to the preliminary results the maximum deviation of the deformed mirror profile from the profile of the approximating paraboloid during symmetrical loading is ±2.5 mm. The rms value of the deviations over the entire surface of the mirror (symmetrical loading) is 1.5 mm. With skew-symmetrical loading the maximum deviation of the deformed mirror profile from the approximating paraboloid is ±3.6 mm in the vertical cross section. The rms value of the deviations over the entire surface of the mirror produced by the total deformations, i.e., symmetrical and skew-symmetrical, is 1.8 mm.

These results on the distortions of the paraboloidal form of the mirror by elastic deformations from the dead weight, according to our assumptions, can be considerably improved since we checked only a small number of variants (six) with different cross sections of the main load-bearing elements of the structure.*

LITERATURE CITED

1. S. Ya. Shifrin, Problems of the Statistical Theory of Antennas [in Russian], Sovetskoe Radio, Moscow (1970).
2. J. Ruze, Proceedings of the Institute of Electrical and Electronics Engineers, British IEE Conference on the Design and Construction of Large Steerable Aerials, London, June, Publ. No. 21.
3. V. N. Ivanov, I. G. Moiseev, and Yu. G. Monin, Izv. Krim. Astrofiz. Obs., 38, 14 (1967).
4. B. H. von Grahe, Die Sterne, 2, 65 (1972).
5. P. D. Kalachev, Preprint Fiz. Inst. Akad. Nauk SSSR, No. 171, (1971).
6. A. F. Bogomolov and B. A. Poperechenko, Radioéktronika, 13, 482 (1970).
7. A. I. Segal', Proekt i Standart, No. 3, 22 (1937).
8. I. E. Livshits, Characteristics of the Calculation of Symmetrical Three-Dimensional Structures [in Russian], Izd. Lit. po Stroit., Moscow (1965), Part 10, p. 273.
9. P. D. Kalachev, Tr. FIAN, 38, 60 (1967).

---

* The Committee on Inventions and Discoveries has resolved to issue an Author's Certificate for a patent on the scheme described here for a mirror system with a parabolic mirror (patent 171283 26-9: Bulletin of Inventions, 1974, 6).

STUDY OF ELASTIC PROPERTIES OF A FULLY STEERABLE
PARABOLIC ANTENNA FOR A RADIO TELESCOPE

P. D. Kalachev and V. E. D'yachkov

The ever more complicated problems which appear before radio astronomy (including long-range radio astronomy) and space radio communication demand the creation of ever more efficient radio telescopes.

At present both in the USSR and abroad there are in operation radio telescopes of various constructions and sizes intended for work in different wavelength ranges. Of these the fully steerable radio telescopes having an antenna in the form of a parabolic mirror comprise the greatest number. While the creation of such radio telescopes is connected both with great difficulties and with great financial expenditures, the serious advantages which they possess stimulate their further development and perfection.

The efficiency of radio telescopes having parabolic antennas is characterized by well-known parameters — the resolving power and the sensitivity. And since both these parameters for a given wavelength are determined by the dimensions of the parabolic mirror, the increase in the efficiency of radio telescopes of this type is connected with the increase in their size.

The principal difficulties which arise in the planning of new large radio telescopes with fully steerable parabolic antennas are, first, the fact that the deformations of the mirror increase in proportion to the square of its diameter (for a given system) and second, the necessity of making it possible to rotate the mirror about the horizontal axis requires of the structural scheme of the mirror system a combination of mutually contradictory conditions in the construction of the mounting. On the one hand, the mirror supports must not obstruct the rotation of the mirror about the horizontal axis, which is satisfied with two supports (in accordance with the two supporting pivots). On the other hand, the mirror supports, i.e., its suspension, must be such that the mirror is deformed radially symmetrically, remaining a solid of revolution, and moreover the deformed mirror must even remain a paraboloid of rotation for which only the focal distance is changed (homologous deformations). The surmounting of these difficulties is required for the attainment of high focusing properties of the mirror and their preservation with any direction of the antenna in space.

According to the Ruze equation, which relates the losses in efficiency of parabolic antennas with the errors in their real reflecting surface, these losses are very sensitive to errors in the mirror surface. The losses in efficiency increase exponentially with an increase in the errors [1]

$$\Delta W = 1 - \exp\left[-(4\pi\varepsilon/\lambda)^2\right], \tag{1}$$

173

where $\bar{\varepsilon}$ is the rms value of the error in the mirror surface (the deviation of the actual surface from the calculated surface) and λ is the wavelength. The value $\bar{\varepsilon}$ includes fabrication errors and errors caused by elastic deformations, including thermoelastic deformations.

In the present article we will take into account only elastic deformations arising from external effects, and of the external effects we will consider only the dead weight and wind.

As already mentioned, the deformations of a mirror are proportional to the square of its diameter [2]:

$$\delta_d = KD^2,$$

where D is the mirror diameter and K is a proportionality coefficient which depends on the number of mirror supports and their arrangement, i.e., on the structural scheme of the mirror system and primarily on the suspension system of the main mirror.

In previous constructions having a two-support suspension of the main mirror the coefficient K has been very large and the mirror, even with symmetrical loading, has deformed in such a way that it ceased to be a paraboloid of rotation. The 76-m parabolic mirror of the radio telescope at Jodrel Bank (England) can serve as an example of this.

As a result of the intensive work of specialists in this area in the last ten years a number of reports have appeared both in our country and abroad devoted to problems of the constrution of parabolic mirrors having deformations of a specific nature [3-6].

The concept of homologous deformations, for which the deformed mirror remains a paraboloid of rotation with a different focal distance, was introduced in [5]. However, cyclic suspensions, including multisupport suspensions for a parabolic mirror, providing for radially symmetrical deformations are required for the realization of such deformations. A very effective element for such suspensions is the multirod pyramid developed at the P. N. Lebedev Institute of Physics of the Academy of Sciences of the USSR [6, 7] and used in the 100-m radio telescope of Bonn University [8] and in plans for other radio telescopes, particularly in the plan for a radio telescope with a 70-m mirror.

The concept of adjusted deformations, which prevent the appearance of defocusing when the mirror is rotated about the horizontal axis, was introduced in [9]. The essence of adjusted deformations consists in the fact that when the antenna is rotated about the horizontal axis the focal distance of the parabolic mirror varies in accordance with the change in the distance from the secondary mirror (or from the exciter) to the apex of the main mirror which takes place because of deformations of the suspension of the exciter system. Adjusted deformations of a mirror system are the most general in the sense of an increase in the efficiency of parabolic antennas, and homologous deformations in our opinion can be considered as a particular case of adjusted deformations when the paraboloidal shape of the deformed mirror is preserved, but there is defocusing of the mirror system when the antenna is rotated in elevation.

In connection with homologous deformations it is necessary to note the following. Usually "homologous structures" for parabolic mirrors are understood to be those structures which not only assure the preservation of the deformed mirror but satisfy this condition for relatively large absolute deformations, i.e., the structure is relatively light and flexible (deformable). In our opinion this conceals a serious deficiency of such parabolic mirror structures if they are not placed within the shelter of a radio-transparent dome. This deficiency consists in the poor resistance to wind effects. It will be shown below that even with a moderately light wind (on the order of 10 m/sec), for which the wind loads are about an order of magnitude less than the dead-weight loads, the deformations from wind loads which distort the paraboloidal shape of the mirror are comparable to and can even exceed those from the dead weight.

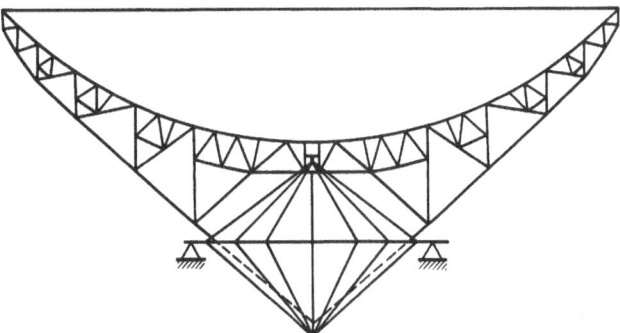

Fig. 1. Structural scheme of a mirror with a cyclic suspension.

The use of complicated servomechanisms for the automatic movement of the exciter (secondary mirror) is not required with adjusted deformations of the mirror system, whereas a homologously deformed parabolic mirror requires devices for the automatic adjustment of the exciter position in order to prevent defocusing.

Thus, the considerable enhancement of the efficiency of parabolic antennas of fully steerable radio telescopes is connected with the necessity of developing and studying new structural schemes for the mirror systems and primarily suspension schemes for the main mirror. On the other hand, the study of optimum schemes from the point of view of providing the fullest realization of adjusted deformations requires the development of the theory of the design calculation on the rigidity of such systems. The final judgment on the quality of scheme selected for the main mirror and its suspension can be had only as a result of a study of the elastic properties of the system, which is connected with multiple static calculations of a large number of variants differing both in the cross sections of the elements of the load-bearing structure and in the mutual arrangement of the supports for the parabolic (main) mirror.

A study of the elastic properties of a mirror system is carried out on a specific structural scheme, for which we have chosen a scheme with a cyclic suspension of the main mirror in the form of a double multirod supporting pyramid (Fig. 1). The double rod pyramid consists of two rod pyramids having a common base while the apices are located on either side of the base. This scheme is the same kind as the scheme presented in [10] (modified scheme).

In the present article we examine theoretical, calculated, and experimental materials with the aim of determining the possibility of obtaining a characteristic* value $h \leq 10^{-4}$-10^{-5} for a radio telescope of large dimensions and particularly for one having an antenna diameter $D = 70$ m.

The work of D'yachkov, in which calculating schemes, the construction of parabolic antennas, and the composition of a calculated load are discussed and calculated loads and the method of their determination as well as different methods of calculating an antenna are examined, is devoted to increasing the accuracy and completeness of the allowance for factors of force and heat effects on an antenna both to assure the correctness of the verifying calculation and for design calculations aimed at the selection of a rational scheme on the optimization of the rigidity characteristics of the structural elements.

_____
*$h = \sigma/D$ is the relative deviation of the actual surface from the calculated surface.

As a result of an analysis of different means of calculating an antenna for rigidity and stability the method of displacements was chosen as the most fully appropriate to the type of supporting structure under consideration. The Ritz method, which for rod systems is equivalent to the method of displacements, is essentially used here.

One of the important advantages of the method of displacements is the fact that it allows one to directly determine the displacements of the nodes of the supporting structure, reliable information on which is the most important and immediate task in the planning of an antenna mirror system.

Coordinate functions are chosen which allow one to seek the components of the displacements not for each node separately but in a certain combination of them. The total number of coordinate functions is equal to triple the number of nodes of a radial girder of the antenna (Fig. 2). The presence of axial symmetry of the antenna allows one to consider it as a cyclic structure [1], thanks to which it becomes possible to simplify the coordinate functions and to represent* the displacements of the nodes of the supporting structure in the form

$$u_{mN} = \mathbf{u} x^u_{mN} f^u_m (\rho_i) \cos N_j \psi,$$
$$v_{mN} = \mathbf{v} x^v_{mN} f^v_m (\rho_i) \sin N_j \psi,$$
$$w_{mN} = \mathbf{w} x^w_{mN} f^w_m (\rho_i) \cos N_j \psi,$$

(2)

where i is the number of nodes of a radial girder; j is the number of radial girders; ψ is the central angle between the planes of neighboring radial girders (Fig. 2); N is the order of the harmonic; m is the number of the series of nodes in the radial direction; $f^k_m(\rho_i)$ is equal to unity when i = m and zero when i ≠ m (k = u, v, w); x^k_{mN} is the generalized displacement of a node of the supporting structure of the antenna in the directions: radial u, tangential v, and parallel to the geometrical axis of the antenna w.

A system of equations is constructed using the equality of the total energy of the system to the sum of the potential energy of the deformations in the work of the external forces:

$$U = \sum_s \frac{N^2_s L_s}{2EF_s} - \sum_{ij} (q^u_{ij} u_{ij} + q^v_{ij} v_{ij} + q^w_{ij} w_{ij}),$$

(3)

where N_s is the force, L the length, and EF_s the rigidity of the rod with the number s; q^u_{ij}, q^v_{ij}, q^w_{ij} are the projections of the force applied to the node ij; u_{ij}, v_{ij}, w_{ij} are the displacements of the node ij in the three indicated directions.

In the case under consideration the equations of the Ritz method have the form

$$\frac{\partial u}{\partial x^u_{mN}} = 0, \quad \frac{\partial u}{\partial x^v_{mN}} = 0, \quad \frac{\partial u}{\partial x^w_{mN}} = 0.$$

(4)

By substituting into Eqs. (4) the values of the displacements (2) and the expression for the total energy (3) one can construct a system of canonical equations and after transformations

---

*In [12] simplified equations are obtained for the determination of the forces in the chord elements of the mirror framework as a cyclic structure (for calculation by the method of forces) and equations for the redistribution of dead-weight loads in the radial elements (skew-symmetrical loading) which considerably simplify the calculation:

$$N_{xi} = \frac{p_0}{\cos \psi/2} \sum_1^i \cos (2_i - 1) \psi/2,$$

$$N_{ri} = 2\wp_0 \sin (2i - 1) \psi/2.$$

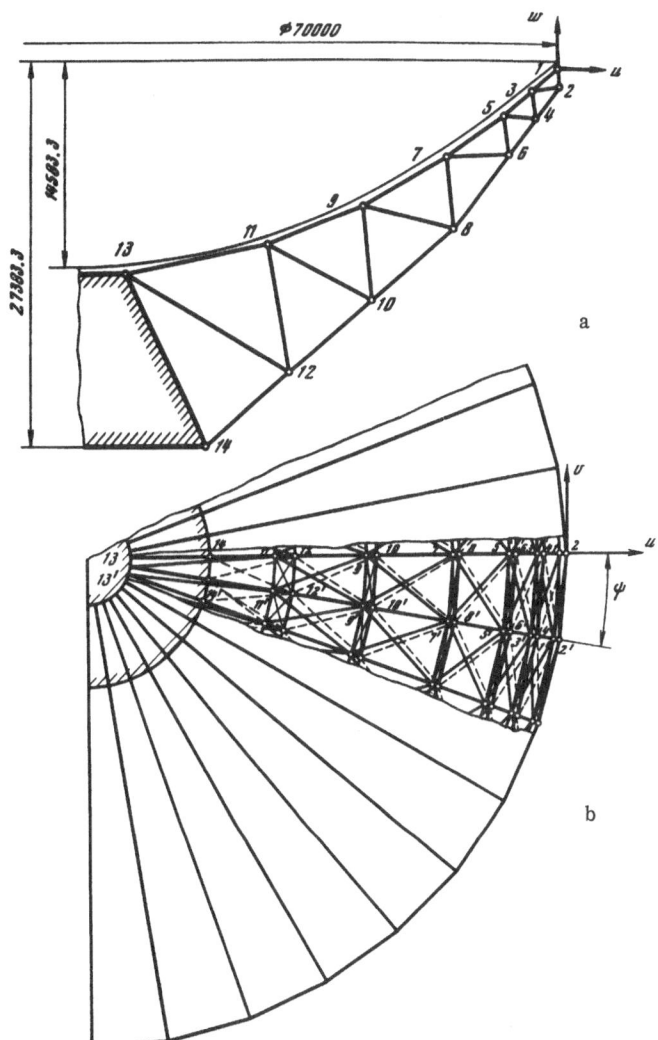

Fig. 2. Diagram of the supporting structure of a parabolic antenna with
D = 70 m. Rods of a radial girder: upper belt: 1-3, 3-5, 5-7, 7-9, 9-11,
11-13; lower belt: 2-4, 4-6, 6-8, 8-10, 10-12, 12-14. Rods of chords
between radial girders: 1-1', 2-2', 3-3', 4-4', 5-5', 6-6', etc. Struts between neighboring radial girders in the planes: a) of the upper belt: 1-3',
1'-3, 3-5', 5-3', 5-7', etc.; b) of the lower belt: 2-4', 2'-4, 4-6', 6-8', etc.

reduce it to the form characteristic for the method of displacements [13]:

$$\sum_{1}^{3} \sum_{mN} x_{mN}^{k} r_{mNM_n}^{k} = R_{mN'}^{k}, \qquad k = u, \ v, \ w. \tag{5}$$

Here in addition to the known designations we introduce new ones: M_n is the number of the equation and R and r are the reactions of the additional braces.

Having determined the matrices of the coefficients $r_{mNM_n}^{k}$ and the values of the free terms R_{mN}^{k} one can, by solving the system of canonical equations (5), find the components of the displacements of the nodes of the supporting structure.

WIND LOADS

Determination of Distribution of Wind Loads

(Overpressure) over Mirror Surface

The dependence of the pressure produced by a wind stream on the surface of a parabolic antenna on the forces and moments relative to the operating aiming axes of the radio telescope is derived through the generalization of experimental data of aerodynamic laboratories on the results of blowing on models of antenna installations. This dependence allows one to determine from the results of blowing on antenna models, with an accuracy characteristic for aerodynamic data and sufficient for practical calculations, the pressures on the surface of a paraboloid on the basis of the data of ordinary blowing with the geometrical axis of the antenna directed both horizontally and toward the zenith.

The materials of experimental studies of a metal antenna model under the conditions of a wind tunnel are presented in the work of V. E. D'yachkov (Leningrad Institute of Railroad Transport), from which one can conclude that with stable heating of the model using an artificial sun to a temperature of $\sim 100°C$ in a calm, when a wind stream with $V \simeq 5$ m/sec is turned on and the effect of the artificial sun is maintained the heating of the rod model is rapidly reduced to a temperature which differs from that of the surrounding medium by no more than $10°C$.

The practical value of this experimental material consists in the fact that one must not sum the wind deformations of the antenna with its maximum temperature deformations. The data of an experimental test of the displacements of the nodes and the stresses in the antenna rods, as well as comparisons of these parameters with those found theoretically, are presented in addition. The error in the measured values established in this case lay in the range of 8-23%. The supporting structure of a radio telescope antenna 70 m in diameter, for which one of the variants of the scheme is presented in Fig. 2, was calculated using the method developed at the Leningrad Institute of Railroad Transport. The scheme analyzed contains 1296 nodes and 3024 rods. The calculation of the coefficients and free terms of the equations was performed on hand key-operated machines of the Iskra type with an accuracy of ten significant digits.

The total displacements of the antenna nodes located in the diametral plane of symmetry of its deformation were found from the equations

$$\begin{aligned}
f_{\beta 0} &= \sqrt{[-u_{i1}^{G+G'} - u_{i0}^{G}]^2 + [w_{i1}^{G+G'} + w_{i0}^{G}]^2}, \\
f_{\beta 0} &= \sqrt{[u_{i1}^{G+G'} - u_{i0}^{G} + u_{i0}^{q}]^2 + [-w_{i1}^{G+G'} + w_{i0}^{G} - w_{i0}^{q}]^2}, \\
f_{\beta 0} &= \sqrt{[-u_{i1}^{G+G'} - u_{i0}^{G} - u_{i0}^{q'}]^2 + [w_{i1}^{G+G'} + w_{i0}^{q'} + w_{i0}^{q'}]^2}, \\
f_{\beta 90} &= \sqrt{[-u_{i1}^{q} - u_{i0}^{q.l.f}]^2 + [w_{i1}^{q} + w_{i0}^{q.l.f}]^2}, \\
f_{\beta 90} &= \sqrt{[u_{i1}^{q} + u_{i0}^{G'} - u_{i0}^{q.l.f}]^2 + [-w_{i1}^{q} - w_{i0}^{G'} + w_{i0}^{q.l.f}]^2}.
\end{aligned} \tag{6}$$

Here $|u; w|_{i1\beta0}^{G+G'}$ are the components of the displacements of the i-th node of an antenna with elevation $\beta = 0$ from the effect of the skew-symmetrical weight load with allowance for icing; $|u; w|_{i0\beta0}^{G}$ are the same from the effect of the axially symmetrical weight load without icing; $|u; w|_{i0\beta0}^{q}$ are the same from the effect of the axially symmetrical wind load with azimuth $\alpha = 0$; $|u; w|_{i0\beta0}^{q'}$ are the same but with $\alpha = 180°$; $|u; w|_{i1\beta90}^{q}$ are the components of the displacements of the i-th node of an antenna with elevation $\beta = 90°$ from the effect of a skew-symmetrical wind load; $|u; w|_{i0\beta90}^{q1.f}$ are the same from the axially symmetrical effect of the lifting force of a wind load; $|u; w|_{i0\beta90}^{G'}$ are the same from the axially symmetrical effect of a weight load of icing.

Here it was assumed that the aligning of the antenna surface took place with the geometrical axis of the antenna directed at the zenith ($\alpha = 0$, $\beta = 90°$).

The largest of the total displacements found from Eqs. (6) was selected, and after adding the facing with the corresponding displacement and allowing for the technological errors in the fabrication of the paraboloid one can take it as the maximum deviation (σ_{max}) of the surface of the deformed antenna from the ideal paraboloid in the determination of the effective area of the antenna from Eq. (1).

A calculation algorithm was constructed for the selection of a rational scheme and optimization of the rigidity characteristics of the elements of supporting structure and the calculation was performed and the antenna deformations were analyzed for different relationships of the rigidity of the rods using an electronic computer such as the M-220. For this the system of canonical equations (5) is represented in matrix form

$$D \cdot x = P. \tag{7}$$

Here D is a matrix of coefficients of equations of z-th order. In the case under consideration (Fig. 1) a radial girder has 12 nodes and consequently $z = 36$; x is a column vector of dimension $z \times 1$, the unknown generalized displacements x_{mN}^{k}; P is a column vector of dimension $z \times 1$, the free terms R_{mN}^{k} of the equations.

Different variants of the schematic solution of the structure illustrated in Fig. 1 were examined, where the coefficients of the equations were determined from the condition

$$D_i = aA + bB + c_1 C^I + c_2 C^{II} + c_3 C^{III} + c_4 C^{IV}. \tag{8}$$

Here

$$a = \frac{EF_{r.g}}{EF_n}; \quad c_1 = \frac{EF_{b.u.b}}{EF_n}; \quad c_3 = \frac{EF_{b.s}}{EF_n};$$

$$b = \frac{EF_{t.r}}{EF_n}; \quad c_2 = \frac{EF_{b.l.b}}{EF_n}; \quad c_4 = \frac{EF_{pl.b}}{EF_n}:$$

A, B, C^I, C^{II}, C^{III}, and C^{IV} are the matrices of the generalized reactions of the rods of the radial girders, tangential rods, and spatial braces joining the neighboring radial braces of the upper and lower belts, struts, and braces, respectively; $E = 2.1 \cdot 10^6$ kg/cm^2 is the modulus of elasticity of the first kind for steel; $F_n = 30$ cm^2 is the cross-sectional area of the rods of the radial girders taken as nominal; $F_{r.g}$, $F_{t.r}$, $F_{b.u.b}$, $F_{b.l.b}$, $F_{b.s}$, and $F_{pl.b}$ are the cross-sectional areas of the rods of the radial girders, tangential rods, and spatial braces joining the neighboring radial braces of the upper and lower belts, struts, and braces, respectively.

The results of the solution of Eqs. (7) on an M-220 computer for several of the variants of the supporting structure schemes examined are presented in the form of graphs in Figs. 3 and 4, which allow one to analyze the effect of the different types of rods on the rigidity characteristics of the antenna.

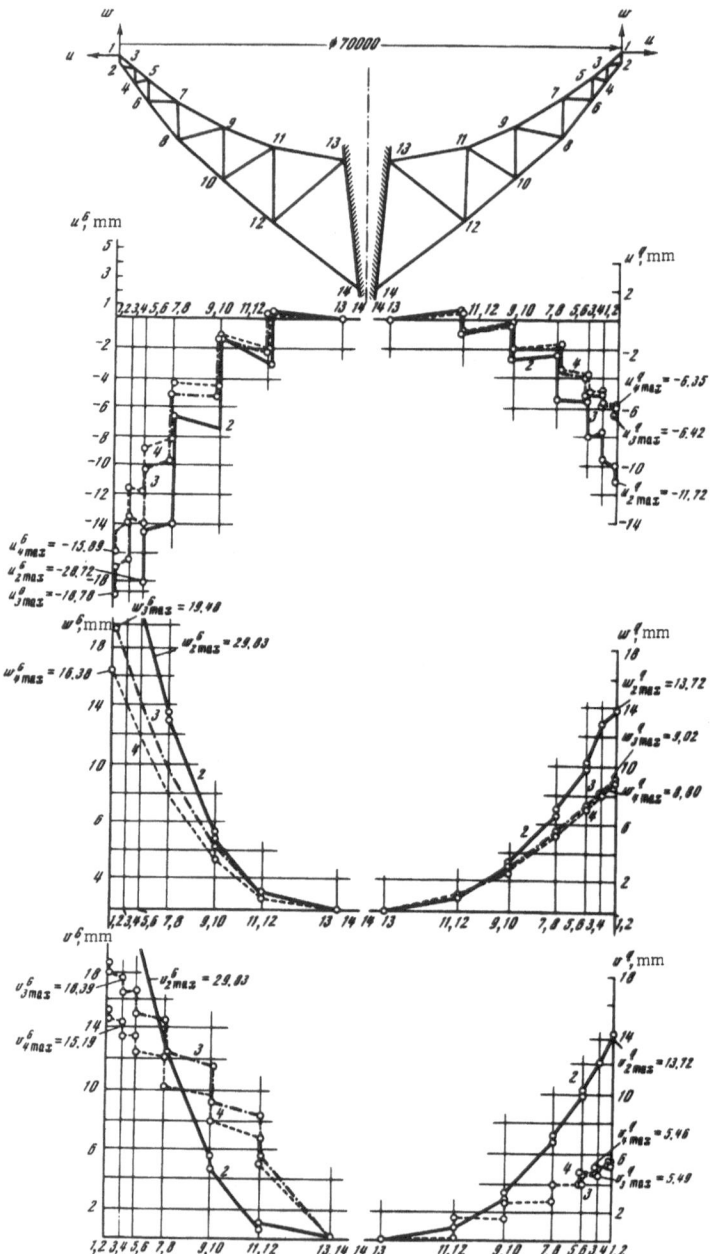

Fig. 3. Axially symmetrical displacements of nodes of an antenna with D =
70 m from the effect of weight loads (with the index G) and wind loads (with
the index q) at a wind velocity V = 18 m/sec.

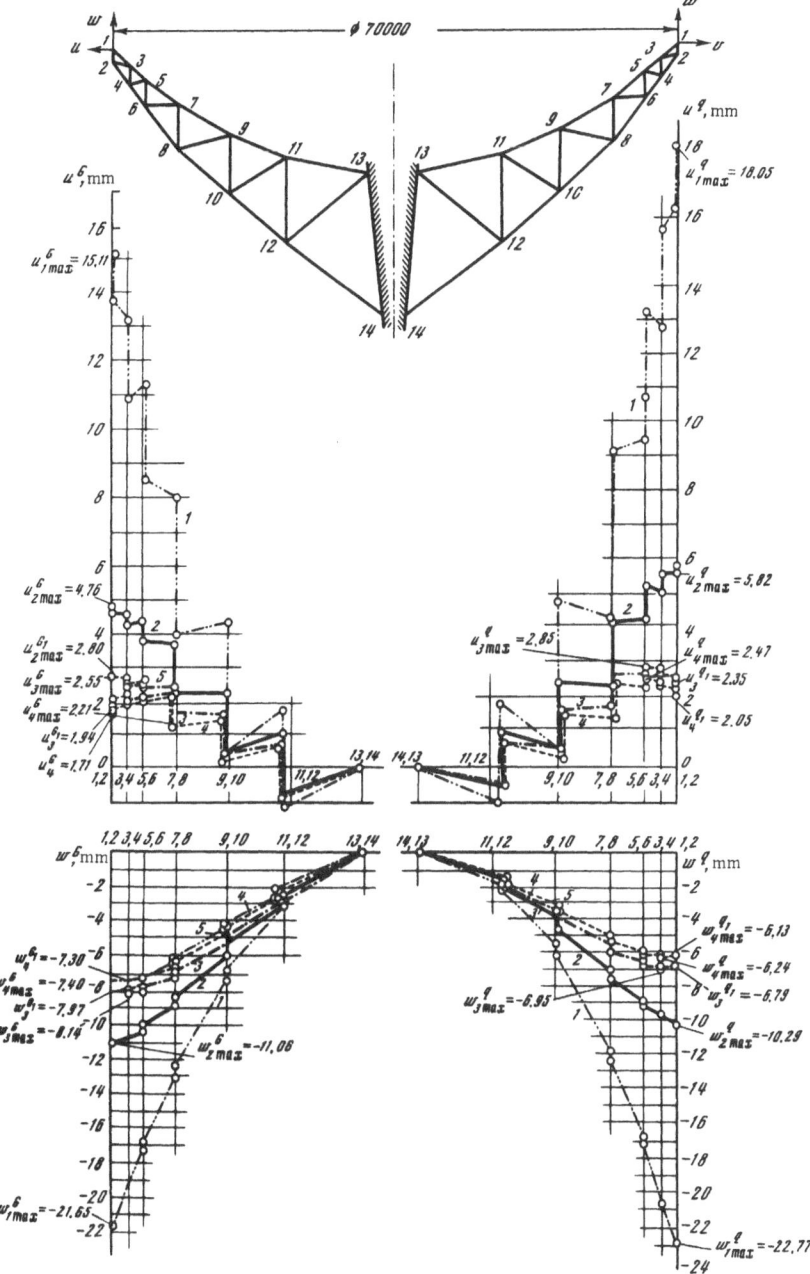

Fig. 4. Skew-symmetrical displacements of nodes on an antenna with D = 70 m from the effect of weight loads and wind loads at a wind velocity V = 18 m/sec.

1. Axially Symmetrical Displacements of Nodes of an Antenna

(Fig. 3) under the Effect of Weight and Wind Loads

a) A comparison of variants 1 and 2 shows that the tangential rods considerably (by two to three times) reduce the antenna deformations, to a greater extent in the direction of the u axis than in the direction of the w axis, both from the effect of weight and wind loads with a relatively small increase in the weight of the supporting structure (see Table 3);

b) The antenna deformations are also reduced by the additional introduction of braces (variants 3 and 4) but to a lesser degree than in paragraph "a," especially if one considers that this makes the structure considerably heavier;

c) Scheme 1 differs from the others in that it is statically determinable, and therefore it has been adopted and sometimes is adopted at present for conducting simplified calculations of antennas, even in the presence of spatial rods for rigidity in addition to the plane radial girders. It is seen from the data presented in Fig. 2 and in Table 3 that the adoption of such simplifications in the calculating scheme is associated at least with the nonuse of large reserves of rigidity and stability of the supporting structure;

d) Variants 2, 3, and 4 are distinguished by the fact that the nodes of the outer two panels (nodes 1-6) have "understated" displacements, as a consequence of which a bend is formed in the edge of the mirror. The presence of such a bend reduces the efficiency of the use of the property of homology of the deformations from the weight load;

e) Variant 5 is characterized by a considerable decrease in the size of the bend indicated in paragraph "d," which is achieved by reducing the rigidity of all the rods of the outer two panels except for the radial girders.

2. Skew-Symmetrical Displacements of Nodes of an Antenna

(Fig. 4) under the Effect of Weight and Wind Loads

a) Scheme 1 is a variable system under the effect of a skew-symmetrical load and therefore it is not considered in this section;

b) The deformations from skew-symmetrical loads are considerably higher than from axially symmetrical loads (by two to three times), especially for scheme 2;

c) The displacements for variant 5 are not presented in Fig. 3 since they differ insignificantly from the displacements of the nodes of scheme 4 for the case of skew-symmetrical loading.

3. Total Displacements of Upper Antenna Nodes (Fig. 5) Located

in the Diametral Plane of Symmetry of Its Deformation under the

Effect of Weight and Wind Loads

a) The graphs presented in Fig. 5 indicate that without the adoption of special measures not one of the schemes examined has a characteristic value $h_c^G \leq 10^{-4}$-10^{-5};

b) Although the absolute displacements from the weight loads are dominant, the displacements which distort the parabolic shape of the mirror, i.e., the displacements relative to an approximating paraboloid, can be made very small through the choice of the optimum structural scheme;

c) Since the displacements in schemes 3 and 4 differ insignificantly one can reduce the number of spatial braces, especially in the peripheral part of the antenna, without marked impairment to the rigidity of the paraboloid.

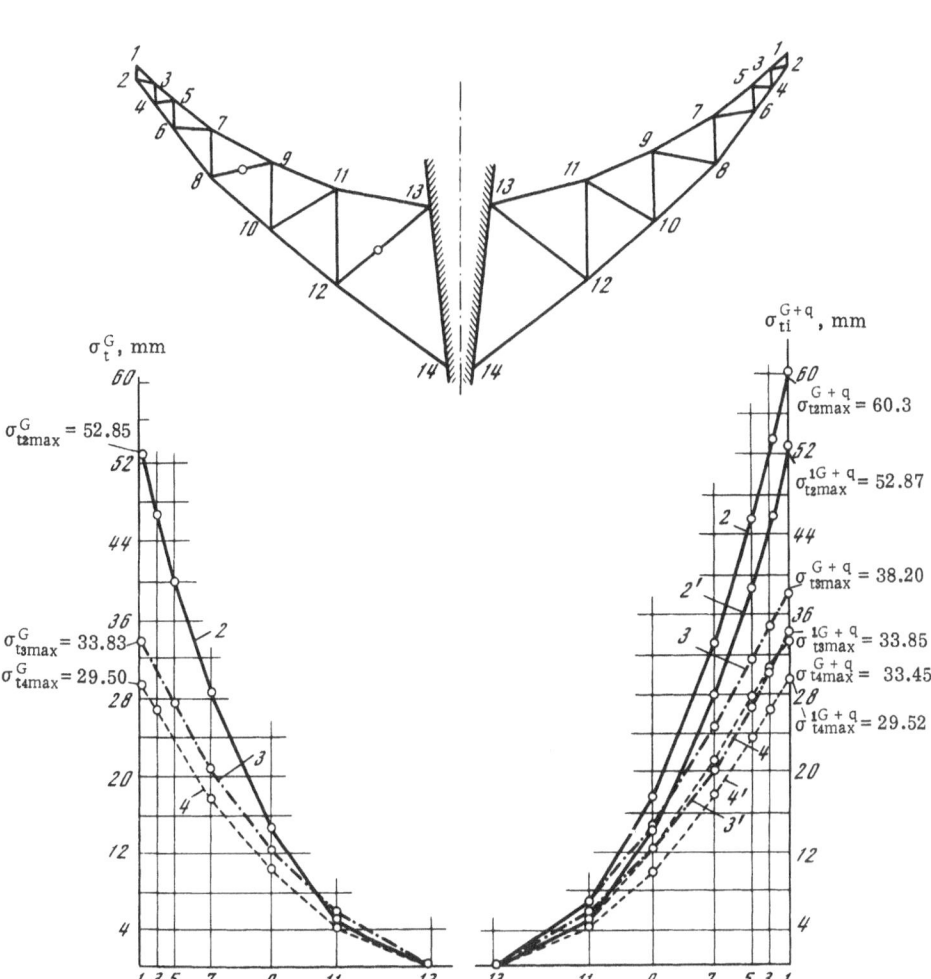

Fig. 5. Total displacements of nodes located in the diametral plane of symmetry of the deformation of an antenna with D = 70 m. σ_{ti}^{G} : total displacements of nodes from weight loads with $\beta = 0, \alpha = 0°$; σ_{ti}^{G+q}: total displacements of nodes from weight and wind loads (V = 18 m/sec); with $\beta = 0$, $\alpha = 180°$; σ_{ti}^{1G+q}: total displacements of nodes from weight and wind loads (V = 10 m/sec) with $\beta = 0$, $\alpha = 180°$.

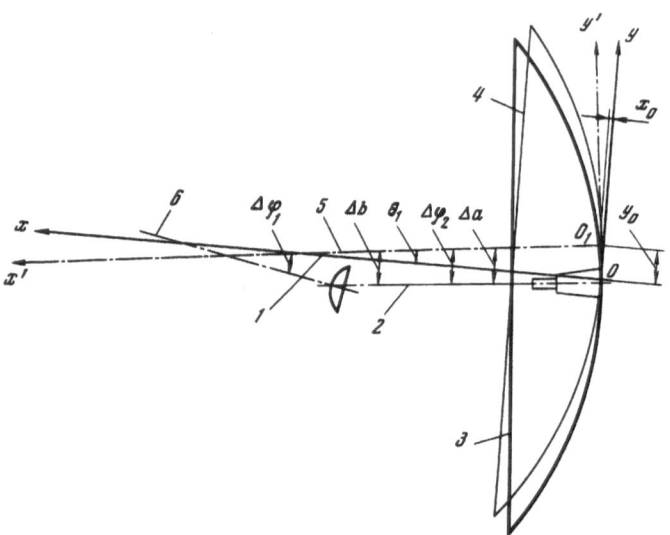

Fig. 6. Diagram of displacement of approximating paraboloid relative
to theoretical paraboloid. 1) Theoretical position of axis; 2) axis of
exciter; 3) approximating paraboloid, F = 20,985.5 mm; 4) theoretical
paraboloid, F = 21,000 mm; 5) focal axis of approximating paraboloid;
6) axis of counter-reflector.

Using the property of homology of the deformations of the mirror surface under the effect of a weight load for the variant of scheme 5 an equation is found for the approximating
paraboloid with a new position for its apex (point 0_1), focal axis 0_1X_1, and focus F = 20,985.5 mm
(Fig. 6).

The basic calculating data for structural schemes 5 and 2 for an antenna 70 m in diameter
are presented in Table 1, from which it is seen that 1) the use of the optimum structural
scheme, allowing one to realize homologous deformations, produced a considerable effect; 2)
scheme 2 has parameters Δb, Δa, etc. which are unsuitable from the point of view of size and
therefore only scheme 5 was examined further.

Two variants of the construction of the approximating paraboloid, which are illustrated
in Figs. 7 and 8 in the form of isodefs, i.e., lines equally distant from the surface of the approximating paraboloid, are examined for structural scheme 5. Here it was assumed that the
distribution of deformations over the entire antenna surface follows Eqs. (2).

Variant I is distinguished by the fact that isodefs close to 1 mm prevail over the greater
part of the surface but the maximum deviation does not exceed 1.75-2 mm.

In variant II (Fig. 8) isodefs close to zero predominate on the working surface but their
maximum reaches 3.5 mm in the zone of low intensity of the radio radiation (close to the edge
of the antenna).

The maximum deviations $\sigma_{\max i}$ of the deformed mirror surface from the approximating
paraboloid produced by different causes are presented in Table 2; their corresponding summation and comparison with the required values are performed. It is seen from the table that

TABLE 1. Basic Calculating Data for an Antenna 70 m in Diameter

Parameter	Designation	Displacement for schemes 5/2		
		from weight	from wind,	m/sec
Maximum deviation of deformed paraboloid from	σ_{max_1}, mm	2.0	—	—
approximating paraboloid		2.0	—	—
Change in focal distance	ΔF, mm	14.5	—	—
		19.4	—	—
Angle of rotation of focal axis of approximating paraboloid	θ_1, ang. min	10	—	—
relative to theoretical axis		17	—	—
Displacement of apex of approximating paraboloid relative	x_0, mm	0.52	—	—
to theoretical value		2.4	—	—
	y_0, mm	106.6	—	—
		195.4	—	—
Maximum deviation of deformed paraboloid from	σ_{max_2}, mm	29.5	2.05	6.6
theoretical paraboloid:		38	4.0	16
$\beta = 0$		0	3.7	12
$\beta = 90°$		0	6.2	25
Transverse displacement of apex of counterreflector	Δb, mm	59	—	—
relative to focal axis of approximating paraboloid		105	—	—
Same for transverse displacement of apex of exciter	Δa, mm	93	—	—
		154	—	—
Same for angular displacement of axis of counterreflector	$\Delta \varphi_1$, ang. min	12.2	—	—
		19.25	—	—
Same for angular displacement of axis of exciter	$\Delta \varphi_2$, ang. min	6.0	—	—
		13.52	—	—
Maximum bending of panel of antenna facing	σ_{max_3}, mm	0.25	0.26	1.0
Total angular rotation of antenna axis owing to deforma-				
tion of its metal structure:				
in elevation:				
$\alpha = \beta = 0$	θ_1, ang. min	10	—	—
		17	—	—
$\beta = 90°$	θ_2, ang. min	—	0.15	0.6
in azimuth:				
$\alpha = 90°$	γ_1, ang. min	—	0.15	0.6
$\beta = 0°$				

the compensatable deformations from the wind load with $\beta = 90°$ and $V = 10$ m/sec are decisive for the efficiency of the antenna system from the aspect of its sensitivity and effective area. Consequently, in this case one should expect a decrease in the value $\Sigma\sigma_{max\ i}^2$ if one uses the aerodynamic effect imparted to the antenna by the introduction of a circular flap mounted with a clearance on the back surface of the reflector. In this case one should keep in mind that with the introduction of this flap the moments on the operating aiming axes of the antenna from the wind load are simultaneously reduced by 1.75 times.*

---

* The study of aerodynamic coefficients at the N. E. Zhukovskii Central Aero-hydrodynamics Institute on the basis of blowing on special models of parabolic mirrors without shields and with circular shields of different widths was carried out by B. A. Garf in 1960-1962 [4]. In this work it was shown that the introduction of circular shields with a relative shield width of 0.07 sharply reduces the aerodynamic coefficients and especially the coefficient of the aerodynamic moment. The physical explanation of this effect consists in the artificial separation of the stream from the leading edge of the mirror.

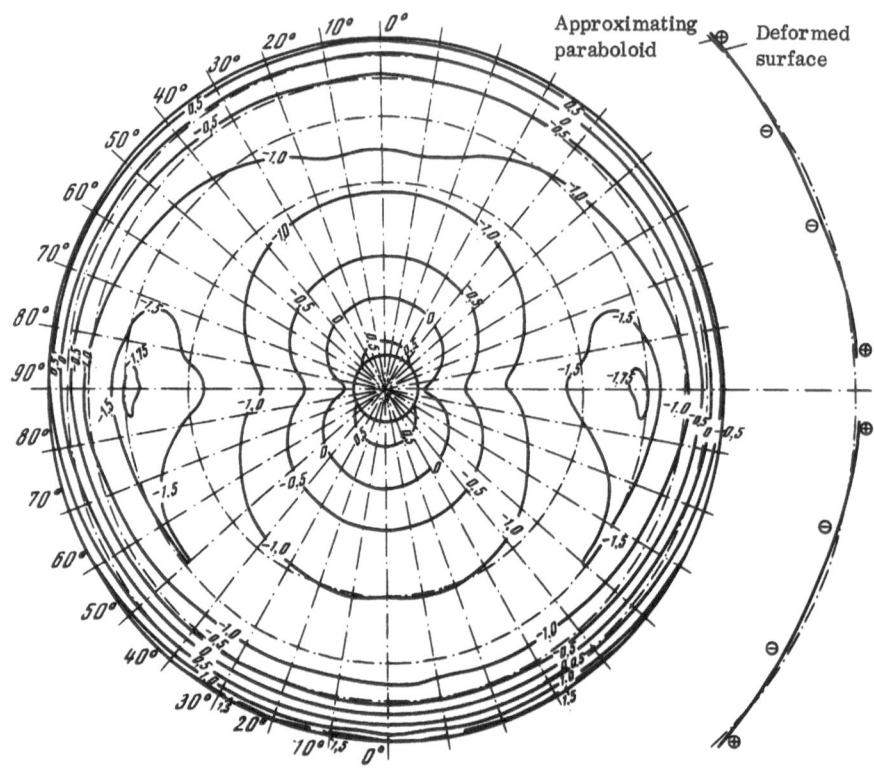

Fig. 7. Isodefs of approximation variant I.

TABLE 2. Maximum Deviations $\sigma_{max\,i}$ of Deformed Mirror Surface from Approximating Paraboloid

Cause of deviation	Deviation in mm with antenna orientation	
	$\alpha = \beta = 0$	$\alpha = 0,\ \beta = 90°$
Weight (relative deformations)	2.0	0
Wind (V = 10 m/sec)	2.05	3.7
Bending of facing (panel) itself:		
from weight	0.21	0.25
from wind V = 10 m/sec	0.26	0.23
Tolerance in fabrication of panel surface	1.0	1.0
Technological accuracy in mounting panels	1.0	1.0
Error in alignment system	1.5	1.5
Maximum dispersion $\Sigma\sigma_{max\,i}^2$:		
in a calm	8.29	4.31
in a wind V = 10 m/sec	12.49	18.07

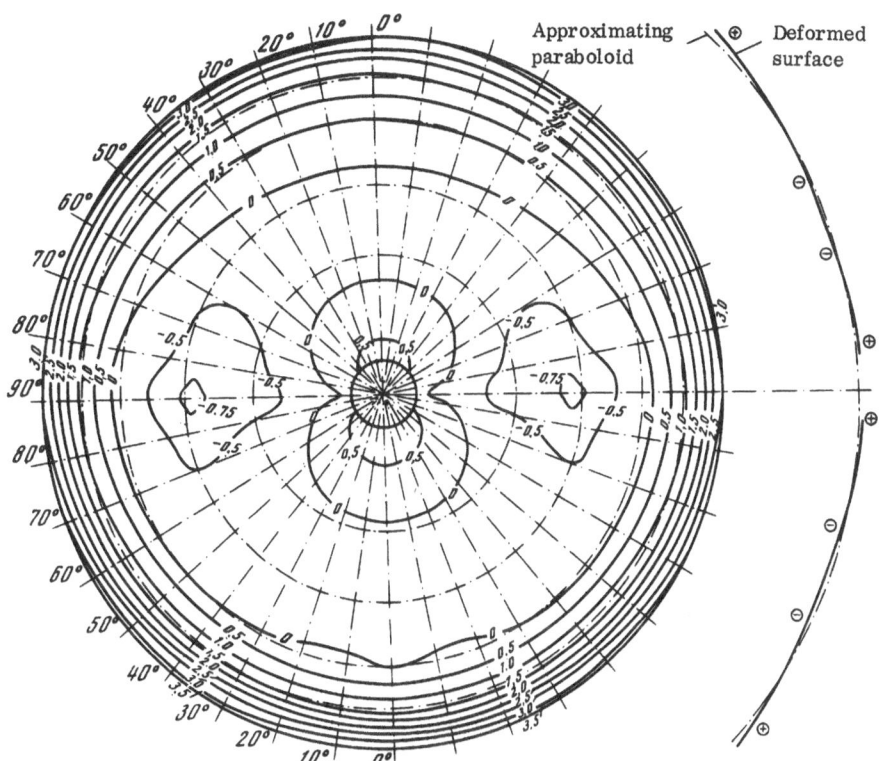

Fig. 8. Isodefs of approximation variant II.

TABLE 3. Characteristic Values h_c, Relative Weights γ, and Other Parameters of Different Variants of Construction Schemes for Antenna 70 m in Diameter

Scheme No.	G_i, m	$\gamma = \dfrac{G_i}{G_i}$	$\sigma^G_{t\,max}$, mm	$\sigma^G_{r\,max}$, mm	Wind velocity V = 10 m/sec					
					$\beta = 90°$	$\beta = 0$	$\beta = 90°$	$\beta = 0$	$\beta = 90°$	$\beta = 0$
					$\sigma^q_{t\,max}$, mm	$\sigma^{G+q}_{r\,max}$, mm	$h_c = \dfrac{\sigma^{G+q}_{r\,max}}{2.6D} \cdot 10^5$		$\gamma h_c \cdot 10^5$	
1	220	0.612	> 40	—	—	—	—	—	—	—
2	255	0.71	38	2	6.2	6	3.4	3.3	2.41	2.34
3	340	0.945	32	2	3.8	4.3	2.1	2.4	2.0	2.3
4	360	1	29.5	2	3.7	4.05	2.03	2.23	2.03	2.23
5	355	0.985	30.0	1.95	3.72	4.0	2.04	2.2	2.0	2.16

The characteristic values h_c, the relative weights γ, and other parameters of the different variants of structural schemes for an antenna 70 m in diameter are presented in Table 3, where the following designations are adopted: $\sigma^G_{t\,max}$ is the maximum total displacement of a node of the supporting structure from the effect of a weight load; $\sigma^G_{r\,max}$ is the same but relative to the approximating paraboloid.

CONCLUSIONS

1. A considerable increase in the efficiency of reflecting radio telescopes with fully steerable parabolic antennas in the form of the classical paraboloid of rotation can be attained through the optimum structural schemes which provide for the realization of adjusted, including homologous, deformations of the mirror system.

2. Our studies showed that while wind loads and the deformations of the main mirror corresponding to them are slight (for mirrors of high rigidity they are about 0.1 of the weight loads at V = 10 m/sec), the deformations caused by them distort the shape of the mirror comparably with the distortion caused by the dead weight.

To decrease the harmful effect of the wind it is necessary, on the one hand, to take measures to reduce the aerodynamic coefficients of the wind loads through the use of circular shields (see [4]) and, on the other hand, to increase the overall rigidity of the main mirror and of the entire system principally through an increase in the number of supports for the mirror.

3. The creation of modern parabolic antennas of high accuracy largely depends on the reliability of the verifying calculation and the depth of analysis in conducting the planning calculation aimed at the selection of a rational scheme and the optimization of the rigidity characteristics of the structural elements.

LITERATURE CITED

1. J. Ruze, Design and Construction of Large Steerable Antennas, Collection of translations (1967), Part II, p. 18.
2. P. D. Kalachev, Tr. FIAN, 38, 72 (1967).
3. P. D. Kalachev, Izv. Vuz. Radiofiz., 6, 399 (1963).
4. H. G. Weiss, MIT Tech. Report, 20, 445 (1968).
5. S. von Hörner, Astron. J., 72, 35 (1967).
6. P. D. Kalachev, Preprint FIAN, A-103 (1962).
7. P. D. Kalachev, Vopr. Radioélektron., Ser. Obshchetekh., No. 11, 29 (1969).
8. O. Hachenberg, Beitrag zer Radioastron., 1, 31 (1968).
9. P. D. Kalachev, Preprint FIAN, No. 171 (1968).
10. P. D. Kalachev, Tr. FIAN, 62, 134 (1972).
11. A. I. Segal', Fundamentals of the Statics and Dynamics of Structures [in Russian], Narkomkhoz RSFSR, Moscow-Leningrad (1938).
12. P. D. Kalachev, Tr. FIAN, 38, 60 (1967).
13. A. P. Filin and A. S. Sokolov, Construction Mechanics for Ships [in Russian], Rechtrans, Leningrad (1957).
14. B. A. Garf, "Mechanisms for the rotation of movable solar installations," Solar Energy [in Russian], Izd-vo Akad. Nauk SSSR, Moscow (1957), p. 62.

AN AUTOMATIC DATA PROCESSING SYSTEM FOR
RADIO ASTRONOMICAL OBSERVATIONS

M. V. Konyukov and V. Yu. Bunakov

INTRODUCTION

The growth in the capabilities of modern electronic computers leads to the fact that the circle of professional users* is continually expanding, while problems which had been considered practically unsolvable can be solved with the expenditure of moderate efforts and means. However, the resort to a computer in the course of a study connected with the solution of complex problems requires the compiling and debugging of programs such that the use of procedurally oriented algorithmic languages and the universal mathematical provision reduces to immaterial the difficulties inherent to programming. Success in the solution of this kind of problem requires the direct participation of the professional user in the programming process, and the time expenditures arising in this connection are so substantial as a rule that the working efficiency of the professional user does not increase and he prefers to reject the use of a computer in the process of conducting the studies [1].

A way out of this situation can be seen in the creation of problem-oriented systems having algorithmic languages capable of providing for programming at the level of the formulations of the problems, avoiding the writing out of problems already formulated in the form of stepwise procedures, the performance of which is characteristic of computers. The creation of this type of system must be preceded by an analysis of the essence of the studies both from the point of view of the extraction of information from the experimental results and from that of its use for the formulation of concepts concerning the subjects studied. The results of this analysis allow one to distinguish the elements in the course of the study for which formalization is possible, to introduce the characterizing experimental and theoretical state of the space, and, finally, to construct the required representation, in other words, to develop the formal bases for a problem-oriented system. The purpose of the present report consists precisely in the creation of the formal bases for a problem-oriented system for professional users working in the field of radio astronomy.

Naturally, following the creation of the formal bases of the system a considerable amount of work by highly qualified programmers is required for the algorithmic and program realization of the system and of the language of intercourse with it, but as soon as this is done the profes-

---

* By the term professional user we understand "scientists, engineers, and all those specialists for whom work on a computer is not the primary profession (beyond a dependence on their familiarity with calculating techniques)" [1].

189

sional users working in the given field can write programs of processing and modeling while having almost none of the concepts of programming in the general sense of the word.

An intuitively defined concept of the system has long existed, but only the appearance of the discipline of system theory, which has developed explosively in recent times, has required the more or less strict definition of the system. Several almost equivalent definitions of the system presently exist and one of them which is convenient for our purposes is presented below: A collection or group of elements needed for the performance of a certain operation is called a system; the word operation denotes an action for the attainment of a definite aim. From this definition of a system it follows that the first step in the development of a system must consist in the more or less strict definition of the aim of radio astronomical studies.

An analysis of the nature of the work performed in radio astronomy leads to the conclusion that the aim of radio astronomical studies can be taken in two senses: In the narrow sense of the word the aim of radio astronomical studies consists in the obtaining from observational data the maximum possible information concerning the electromagnetic field produced at the standard surface of the source of radio radiation*; in the broad sense of the word the aim of radio astronomical studies comes down to the determination of the physical properties of the object from the observational data.

The electromagnetic field studied by radio astronomical means arises as a result of the "illumination" by the astronomical object of a surface whose position and shape are determined by the object and the instruments used. In the creation of a data processing system for radio astronomical observations it seems appropriate to introduce a single standard surface for all the observations of the object under consideration, which is possible if there exist mutually equivalent and mutually continuous transformations of the fields at the observation surfaces into the corresponding fields at the standard surface.†

According to modern concepts the sources of radio radiation subject to study are such that the state of the matter in them can be described only statistically, and the characteristic times of the elementary acts of emission and absorption are considerably shorter than the time constants of the instruments used. Therefore the mutual coherence functions of different orders introduced in optics [2-4] can be used as the values characterizing the radiated electromagnetic field. Naturally, with such an approach one cannot obtain from the observational data an answer concerning the exact distribution of the electromagnetic field because the mutual coherence functions carry information only on the statistical properties of the radiation which has arrived.

Under conditions where the antenna aperture is small compared with the characteristic size of a region of uniformity of the field the radio telescope can be considered as a probe which reacts to the electromagnetic field at a point,‡ and the obtaining of information on mutual coherence functions of different orders requires simultaneous observations at spatially sepa-

---

*In the framework of the proposed system the discovery of fundamentally new sources belongs among the unformalized elements in the process of radio astronomical studies.

†The introduction of a standard surface is possible only if there is a medium with known properties in the region of space between it and the observation surface. It should also be kept in mind that difficulties may arise when a single standard surface is introduced for a source which occupies a considerable part of the celestial sphere.

‡Since the directional diagrams of the telescopes are not δ-functions, the obtaining of information on the field arriving at a point where an antenna with a fixed direction is located entails the problem of reconstruction [5, 6].

rated radio telescopes.* This means that a system intended for achieving the aim in the narrow sense of the word must provide for that data processing, for observations both on individual and on spatially separated antennas, with which one extracts the maximum possible information on the mutual coherence functions of all possible orders.

The question of the system in the broad sense of the word is more complicated. At present the absence of integral equations permitting a unique solution and relating the mutual coherence functions at the standard surface with the physical properties of the matter and fields in the object studied seems to be an indisputable fact (the same is true for the electromagnetic field). This is connected with the fact that the information on the mutual coherence functions can be obtained only for a limited region of space in the vicinity of the earth.† The most important consequence of the property of integral equations indicated above is that a system for attaining the aim in the broad sense of the word can be constructed only on the basis of modeling of the conditions in the astronomical objects.

1. FORMAL BASES OF THE SYSTEM

The formal bases of a system of automatic data analysis for radio astronomical observations contain a number of concepts introduced using the properties of the discipline discussed in the preceding section. These concepts are presented and briefly discussed below.

The Set of Observational Data. The data of observations of objects by radio astronomical means form a set. This set is not empty if at least one observation is conducted the result of which has the following structure: the name of the object; the filter used to conduct the measurements (the filter is understood to be the representation, equivalent to the installation, consisting of the radio telescope, the receiving apparatus, and the recording devices, and the means of processing the recorded signals to obtain the observational results — the actual numbers — and the degree of their reliability); the actual number — strictly the result of the observations; the collection of numbers characterizing the degree of reliability of the observational result (in the simplest case this is the measurement error).

Structure of the Observational Data on the Object. The subset of the set of observational data, consisting of the elements pertaining to the object under examination and adapted to the standard surface and the equivalent collection of filters, forms the basis for forming the structure of the observational data on the object.‡ The structure of the observational data arises as a result of the partition of the indicated subset into classes of fully ordered subsets. The

---

*It must be kept in mind that the information on mutual coherence functions of higher order is fundamentally new and cannot be obtained from functions of lower order. In the simplest case this means that it is impossible to obtain first-order mutual coherence functions from measurements on an individual antenna no matter how narrow the directional diagram.

†The equations relating the properties of the matter and fields in the objects with the fields produced by them at the standard surface are integral equations, but, as it is easy to establish, they do not have a unique solution.

‡The adaptation of elements of the indicated subset consists in the conversion of the results of the real observations to the results of observations of the object using an equivalent collection of filters adjusted to the standard surface. Both the standard surface and the equivalent collection of filters are, generally speaking, different for different objects.

equivalence ratios* appropriate for the given object and the ordering criteria† inherent to each class comprise the basis for the partition [7]. The structure of the observational data is a dynamic formation and represents the complete characterization of the observational data for the object at the moment of formation of the structure.‡

Set of Structures of Observational Data. The structure of the observational data for any object is an element of the set of structures of observational data. It arises from the set of observational data through the formation of the structures and determines the experimental state of the study of astronomical objects by radio astronomical means.

Informational Structure of the Object. To construct the informational structure of the object one must know: the set of filters, adjusted to the standard surface, of the structure of the observational data for this object; the distribution of the field from the same object which is possible in the set. The effect of the equivalent collection of filters on the allowable field adjusted to the standard surface leads to the appearance of two sets. One of them, analogous to the set of observational results of the structure of the observational data on the object, consists of the actual numbers and represents the intrinsic informational structure of the object, while the other is analogous to the set of vectors characterizing the degree of reliability of the observational results. The sets generated have the same ordering as the corresponding sets of the structure of the observational data on the object.

The Space of Informational Structures of the Object.: The fields at the standard surface which are possible for the object under consideration form a metric space.** The action on each element of this space by the mapping which leads to the formation of the informational structure generates the metric space of informational structures of the object.†† As the distance between points of the space of informational structures one can take [7]

$$\rho(x, y) = \max_{1 \leqslant i \leqslant n} |x_i - y_i|,$$

where n is the number of components of the informational structure of the object while x_i and y_i are the suitably numbered components of the informational structures x and y.

---

*The choice of the binary equivalence ratios needed for the partition into classes is determined by the collection of filters and the peculiarities of the object under consideration.

†The structure of the observational data actually consists of three ordered sets: the set of filters adjusted to the standard surface; the set of observational results; the set of vectors characterizing the degree of reliability of the observational results. The term "structure of the observational data" will often be used below to denote only the set of observational results.

‡For each element of the structure of the observational data for the object one must indicate its connection with the mutual coherence functions of the field produced by this object at the standard surface. The structure of the observational data which has been introduced is easily generalized to nonstationary objects.

**By considering the fields as belonging to a class of functions having derivatives up to second order inclusively one can take as the distances between points of the space of the expected fields [8] the value

$$\rho(f_k, h_k) = \sum_{i=0}^{2} \max |D_i f_k - D_i h_k| \qquad \text{for} \quad z \in Z_0,$$

where f_k and h_k are functions of the arguments of z determining the field; Z_0 is the set of values of the arguments yielding the field at the standard surface; D_i are the operators of the derivatives of i-th order.

††The mapping of the points of the space of possible fields onto the space of informational structures is equivalent and continuous but mutual equivalence and mutual continuity are absent for it.

Suppose X is the structure of the observational data on an object, adjusted to a certain standard surface and equivalent collection of filters $\hat{\Phi}$, while ε is a point of the space of possible fields of the object at the standard surface. Then the mapping operation $\hat{\Phi}$ at ε gives a point of the space of informational structures of the object, while the equation

$$\hat{\Phi}(\varepsilon) = X \tag{1}$$

determines those points of the space of possible fields whose informational structure coincides with the structure of the observational data on the object.

In solving Eq. (1) one can use the method of parameters for the assignment of the points of the space of possible fields. In this case the solution of Eq. (1) can sometimes be reduced to a variational problem of a search for parameters, and with the appropriate metrics in the space of informational structures it can be reduced to the method of least squares. However, one must remember the limitations inherent in a search for solutions using the method of parameters, and especially the determining role of the a priori information.

Informational Equations. The equations which allow one to obtain the distribution of the field from the object under consideration at the standard surface are called informational equations. The system of equations of electrodynamics, the parameters of which are the functions determined by the properties of the medium both in the region of generation and in the region of propagation of the radiation, can be used as the informational equations in the study of objects by radio astronomical means. The informational equations allow one to construct the space of possible fields at the standard surface from the space of parameters characterizing the object.

Under conditions where the equivalent collection of filters acts on the distribution of mutual coherence functions at the standard surface, the generalized equations of propagation of the mutual coherence functions of the appropriate orders can be taken in a natural way as the informational equations. For example, in the simplest case of a collection of filters for which the intensity distribution at the standard surface serves as the initial information one can take the radiation transfer equation as the informational equation.

Modeling Operator for the Object. The modeling operator for the object is considered as assigned if in accordance with the concepts adopted concerning the physical conditions in it:

a) a set of functions is determined for which the differential equations used in the operator have meaning;

b) the boundary conditions and initial conditions distinguishing the subset of functions, called the region of definition of the operator, are assigned;

c) the differential equations acting on the elements of the region of definition of the operator are assigned;

d) the region of values of the operator is determined [9].

When the modeling operator is known the construction of the model comes down to the search for functions from the region of definition for which the operator takes on the assigned values (functions from the region of values of the operator).*

---

* Depending on the concepts concerning the physical conditions in the object the operators used can be integral and integro-differential operators. The informational equations enter into the system of equations of the model and sometimes play an essential role. Their distinction, however, is that the solution of the informational equations must be obtained from the standard surface, whereas the solution of the other equations of the model is sought in the region of space occupied by the object under consideration.

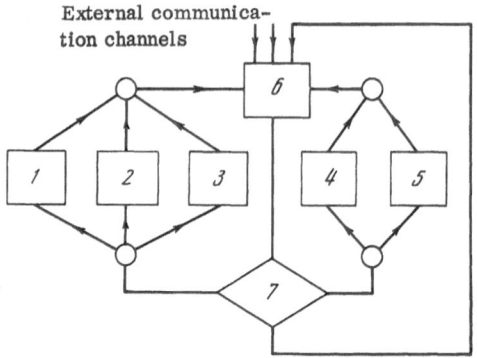

Fig. 1. Schematic diagram of the system. 1) Generation of newly introduced filter; 2) planning of experiment (generation of parameters of the experimental apparatus); 3) formation of structure of observational data; 4) search for an element of the space of modeling operators; 5) generation of new elements of the space of modeling operators; 6) system of automatic data processing for radio astronomical observations; 7) unit for accepting the solution.

Space of Modeling Operators of the Object. The variations in the modeling operators compatible with the concepts concerning the physical conditions in the object generate a set. [*] The introduction of metrics into the set of modeling operators which has been generated converts this set into a space of modeling operators of the object.

The existence of a continuous mapping onto the space of informational structures is assumed for each point of the space of modeling operators. Sometimes it is convenient to represent this mapping as a composite of two mappings: the mapping of the points of the space of modeling operators of the object onto the space of its possible states and the mapping of the points of the space of possible states onto the space of informational structures.

Suppose $\hat{G}$ is the mapping of the points of the space L of modeling operators of the object onto the space of possible fields at the standard surface. Then the action of the composite $\hat{\Phi} \circ \hat{G}$ of mappings on L gives a point of the space of informational structures of the object, and the equation

$$\hat{\Phi} \circ \hat{G}(L) = X \tag{2}$$

determines those points of the space of modeling operators of the object for which the informational structure coincides with the structure of the observational data on the object. [†]

---

[*] The indicated variations include variations in the differential equations as well as in the regions of definition and the values of the operator.

[†] The space of modeling operators of the object is the region of definition, while the space of informational structures is the region of values of the mapping $\hat{\Phi} \circ \hat{G}$. We note that phenomenological modeling, in which the space of modeling operators of the object is replaced by the space of its possible states, is common in the practice of radio astronomical studies. The modeling equation (2) preserves the form but $\hat{G}$ and L have different meanings.

In the solution of Eq. (2) one can use the parameter method to assign the points of the space of modeling operators. In this case the solution of Eq. (2) can sometimes be reduced to a variational problem of a search for parameters. However, one must remember the limitations inherent in a search for solutions using the method of parameters, and especially the determining role of the a priori information.

2. SCHEMATIC DIAGRAM OF SYSTEM

A system of automatic observational data processing can be developed to attain the aim of radio astronomical studies. A schematic diagram of a system of this kind is presented in Fig. 1. It consists primarily of two circuits: a circuit for the formation of the structure of the observational data, which in addition to the intrinsic formation of the structure of the observational data provides for the generation of new filters and the planning of the experiment, and a circuit for searching for an element of the space of modeling operators of the object (the possibility of expansion of the space of modeling operators of the object through the generation of new elements is provided for in it).

In addition to the indicated circuits the system includes channels for communication with the sources of information, providing for its reception from the literature.

Since the organization of the search for elements of the space of modeling operators requires the use of methods of heuristic programming, the system includes a unit for accepting the solution and control channels allowing one to accomplish entry into the system, bypassing its main circuits.*

CONCLUSION

The creation of a system of automatic observational data processing requires the solutions of two main problems: 1) the organization of the collection, sorting, and storage of the information supplied from the experimental installations†; 2) the creation of a library of programs for processing the observational data, for modeling, and for organizing the intercourse between the investigators and the programs of the library and the stored information, both that subject to processing (the observational data processing proper) and the processed data (the modeling).

The first is mainly an engineering problem: Memory devices for the information must be selected, the means for its transmission from the experimental installations to the memories must be developed, algorithms for sorting the information supplied must be constructed and realized in the form of programs, and principles for the archive storage of the information must be developed.

The second problem actually comes down to the creation of the mathematical provision: Methods must be developed for solving the problems of processing and modeling, algorithms constructed for the solution of these problems, the corresponding program libraries created, and finally, a language must be chosen or developed for communication with the stored information and the libraries. Various means for the creation of the mathematical provision are conceivable or with their determining influence the algorithms are developed and the programs are created for the concrete problems of processing and modeling, and the access to the program libraries and the stored information is conducted in the universal lan-

---

*In the presence of several objects the system of automatic observational data processing will consist of subsystems of the indicated type.

†The volume of information supplied from the experimental installations in radio astronomy and the algorithms for its processing with the current computer costs make it unfeasible in practice to operate the system in real time [10].

guage of communication with the computer. There are a number of disadvantages inherent in this course, and among them the most important are the following: A reasonable generality is lacking, generally speaking, from the library programs of processing and modeling, which often creates difficulties in the use of library programs to solve problems which are close in meaning; the creation of large processing and modeling programs requires expenditures of time which are unacceptable from the point of view of the professional user [11, 12].

An undoubted advantage of this course is the relatively short time needed to create the necessary libraries with a relatively low qualification of the users as programmers and the absence of a need to develop a language of communication with the computer.

A basically different course is the systems approach: One must develop a single complex of programs, providing for the attainment of the aim of the radio astronomical studies, in the use of which the professional user spends a sufficiently short time in programming. As the guiding idea in the development of the bases for such a complex we used the opinion of the authors of the SIMULA-67 language concerning future languages of communication with the computer: "The future evidently belongs to the 'specialized language,' which in all probability will be extremely specialized. Such languages can contain the basic concepts and methods related to the given field. They permit the user to formulate concrete problems on the basis of his previous experience and knowledge. At the same time it is necessary that the languages themselves be capable of expansion in the sense that all new information in the given field of study can easily be incorporated into the language by the investigator himself" [12]. It seems to us, however, that the realization of this idea must proceed along the course of the creation of problem-oriented systems of observational data processing (all the special features of the discipline and the capacity for supplementation by the user will be incorporated in just these systems) and the development of a language of communication with the system which is simple enough in grammatical structure.*

It is not hard to see that the algorithmic and programmed realization of a problem-oriented system requires considerable expenditures of the work of highly qualified programmers, and it is just this which can have a restraining effect on the development of the systems approach to the development of the mathematical provision for work in the field of radio astronomy.

In conclusion the authors consider it their pleasant duty to thank I. M. Dagkesamanskaya, R. D. Dagkesamanskii, and V. A. Udal'tsov, who provided support for the work at various stages of its completion.

LITERATURE CITED

1. A. Scherr, Analysis of Time-Shared Computer Systems, MIT Press (1967).
2. A. Blanc-Lapierre and P. Dumontet, Rev. Opt., 34, 1 (1955).
3. E. Wolf, Proc. Phys. Soc., A230, 246 (1955).
4. M. Born and E. Wolf, Fundamentals of Optics [Russian translation], Nauka, Moscow (1970).
5. R. N. Bracwell, Proc. IRE, 46, 106 (1958).

---

* The creation of a problem-oriented system of observational data processing requires libraries of processing and modeling programs providing for the operation of the system. The distinguishing feature, however, is that they, as a rule, will be the generators of programs for the solution of problems of a specific class. It must be pointed out that both the distinguishing of the classes and the programmed realization of the appropriate generators may require the surmounting of considerable difficulties.

6. A. N. Tikhonov, V. V. Vitkevich, V. S. Artyukh, V. B. Glasko, A. V. Goncharskii, and A. G. Yagola, Astron. Zh., 46, 472 (1969).

7. A. N. Kolmogorov and S. V. Fomin, Elements of the Theory of Functions and Functional Analysis [in Russian], Nauka, Moscow (1968).

8. I. M. Gel'fand and S. V. Fomin, Variational Calculation [in Russian], Fizmatgiz, Moscow (1961).

9. M. A. Neimark, Linear Differential Operators [in Russian], Gostekhizdat, Moscow (1954).

10. B. G. Clark, Ann. Astron. and Astrophys., 8, 115 (1970).

11. H. Markovitz, B. Hausner, and J. Carr, SIMSCRIPT: Simulation Programming Language, Prentice-Hall (1963).

12. W. Dahl, B. Moorhouse, and Newgard, SIMULA-67, a Universal Language for Programming [Russian translation], Mir, Moscow (1969).

THE SYNCHRONOUS-TRACKING DRIVE SYSTEM FOR THE RTI-7.5/250 RADIO TELESCOPE OF THE MOSCOW TECHNICAL COLLEGE

A. A. Parshchikov and I. A. Emel'yanov

To achieve the designed accuracy of the antenna of the RTI-7.5/250 of the Moscow Technical College (MTC) the minimum width of the directional diagram at $\lambda = 1$ mm had to be $\theta_{0.5} \approx 30"$. For practical work on radio telescopes it is necessary to know the position of the directional diagram of the antenna in space by no worse than $(1/3-1/6)\theta_{0.5}$, i.e., by 10-5". Such high accuracy demanded the development of a specialized control system which would make it possible not only to achieve the accuracy in tracking but also had the required modes of control for the solution of a wide circle of problems.

The plan of a synchronous-tracking drive (STD) system was developed at the MTC with the participation of colleagues at the P. N. Lebedev Institute of Physics of the Academy of Sciences of the USSR. The realization of the plan allows one to have the following modes of operation on RTI-7.5/250 antennas of the MTC: automatic aiming in accordance with the data of a digital computer; semiautomatic slow-motion aiming; semiautomatic rapid-motion aiming; scanning; readjustment of the antennas in the entire range of working angles; functional control of the entire control complex in the mode of local control; automatic aiming in accordance with the data of a photoguide.

The choice of the required modes of operation predetermined the use of a two-motor system with a mechanical differential (Fig. 1). Direct current motors of the MI-12 and DPM-11 types, which have a number of advantages of a regulatory nature compared with alternating current motors, were chosen as the actuators. Amplidynes of the ÉMU-3A and ÉMU-25 A_3 types, respectively, having specially matched control windings were chosen as the power amplifiers for the motors used.

The accuracy and stability of the control system, which was treated as a combined linear impulse system, were studied for the modes of operation of the radio telescope selected. The required frequency of working operations on the system and the coefficients of stability in amplitude and phase are determined from the results of these studies.

In highly accurate automatic control systems whose errors are measured in seconds the main difficulties in the stabilization of the system are caused by the presence of play in the power-reducing gears, since in this case even a slight amount of free play proves to be comparable with the dynamic error of the system.

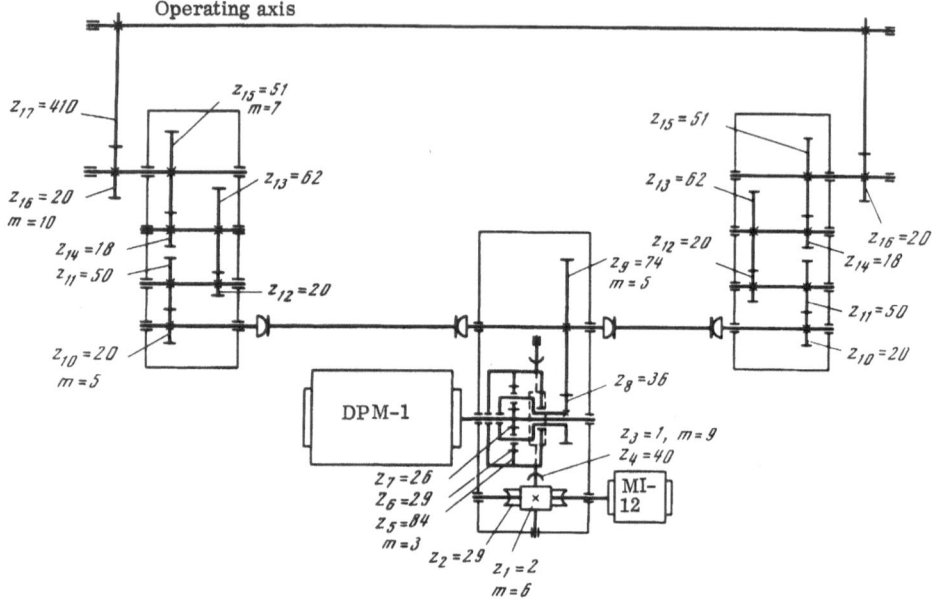

Fig. 1. Kinematic diagram of mechanical drive in elevation.

A number of construction measures are taken for the partial elimination of the free play in the gears:

a) Eccentric bushings which regulate the center-to-center distances are introduced.

b) The driving (main) gears are pressed against the rotation sectors (the gear of the supporting-rotating mechanism). In addition, the construction of the mechanical elevation drive incorporates the possibility of introducing a preliminary tightness with a one-sided choice of the clearance in the gears due to the unbalance of the mirror and the intermediate structure, on the one hand, and of the rotation sectors with the counterweight, on the other.

TABLE 1

Characterizing parameter	Azimuth	Elevation (zenith distance)
Range of working angles	±172.5°	2.5–90°
Velocity of aiming by computer data	0–205''/sec	0–325''/sec
Velocity of semiautomatic control	0–205''/sec	0–325''/sec
slow motion	5°/sec	8°/sec
fast motion	10''	10''
Error of aiming by computer data *	10°/sec²	10°/sec²
Maximum possible accelerations	20''.5/sec²	32''/sec²
Accelerations of aiming by computer		

*From measurement of the error signal.

Fig. 2. Telescope control panel.

Fig. 3. The Dnepr digital computer.

In order to neutralize the effect of the free play on the stability of the system, in addition to decreasing it through the construction of the reducers, correcting chains were used which generate signals proportional to the first derivative of the control effect. The main technical characteristics of the RTI-7.5/250 STD of the MTC are presented in Table 1.

The main control panel (Fig. 2) is located between the antennas in the instrument room, where the Dnepr digital computer incorporated into the STD system of the RTI-7.5/250 is also located (Figs. 3 and 4). The main elements of the power drive, the amplidynes, the protection for the drive motors, and the power supplies, are mounted in a separate house in the immediate vicinity of the instrument room.

"Automatic" operation by signals from the digital computer is assumed to be the main mode of operation. The strict limitation on the sizes of the allowable errors imposes a number of demands on the RTI-7.5/250 control system in the automatic mode, namely:

a) the errors in the calculation of the horizontal coordinates of the objects studied and in the law of scanning of the antenna beam in the real ephemeride time scale with compensation for systematic errors caused by the residual inaccuracy in aligning the radio telescope axes must not exceed 1″;

Fig. 4. Inner view of apparatus room.

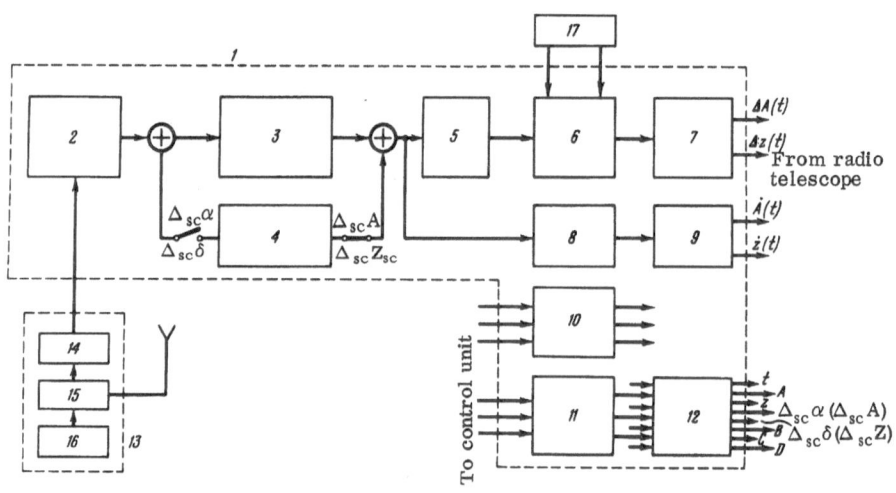

Fig. 5. System of automation of observations on the RTI-7.5/250 of the MTC. 1) Algorithm for automatic mode of observations (Dnepr computer); 2) quadratic interpolation of equatorial coordinates of object and calculation of local stellar time (0.5 Hz); 3) calculation of horizontal coordinates of object with compensation for errors in antenna alignment (0.5 Hz); 4) calculation of scanning law (0.5 Hz); 5) linear interpolation of horizontal coordinates of object (10 Hz); 6) calculation of codes for control operations (10 Hz); 7) digital–analog conversion (10 Hz); 8) calculation of rates of change of horizontal coordinates of object (0.5 Hz); 9) digital–analog conversion (0.5 Hz); 10) printer (0.2 Hz); 11) analog-digital-analog con- Hz); 12) PL-80 tape punch; 13) timer; 14) pulse shaper; 15) frequency–comparison device (Ch1-29); 16) Ch5-6 high-stability quartz generator; 17) angle detectors on telescope axes.

b) the accuracy in the determination of the angular positions of the telescope axes must be on the same order;

c) the operating part of the control system must provide the possibility of working out the control operations with the required accuracy.

A system of automation of observations on the RTI-7.5/250 which satisfies the demands listed above was developed and put into operation on the basis of the Dnepr digital computer at the MTC with the participation of the Ukravtomatika Trust (Fig. 5). In the mode of automatic observation of the objects studied the Dnepr computer (Fig. 3) performs the following functions:

1. Calculates with a frequency of 10 Hz in the real scale of ephemeride time t the horizontal coordinates of the observed object, the rates of their variation, and when necessary the scanning law of the radio telescope beam in equatorial (α, δ) or horizontal coordinates (A, z), containing the calculated values of the azimuth A, and zenith distance z, and their derivatives $\dot{A}$ and $\dot{z}$ corrected for the residual inaccuracy in the radio telescope alignment.

2. Determines the codes of the control operations ΔA and Δz as the differences between the calculated and actual positions of the telescope axes and gives the voltages corresponding to them for the object with a frequency of 10 Hz and the voltages proportional to the rates of change of the horizontal coordinates of the object with a frequency of 0.5 Hz.

3. Records the current values of the following parameters on a PL-80 or a printer: the ephemeride time, the corresponding horizontal coordinates of the object, and scanning parameters, and the current parameters of the signal at the output of the radiometer.

The flexibility of the algorithm developed allows one to change from one mode of tracking or scanning to another by a simple change in a small number of constants entered into certain cells of the machine memory. In order to reduce the dynamic error the operating motors on the telescope axes are controlled in position and in velocity. One-channel, 18-place digital pickups whose error does not exceed $2''.5$ are used as the detectors of the position angles of the telescope axes. A timer, constructed on the basis of the comparison of the frequency of a local high-stability generator with the signals put out by a time service, provides the required accuracy of the local time scale with a large margin even with rather infrequent comparison of the clocks (less than once a month).

The main difficulty which arose in putting into operation the system developed was the complexity in matching the required calculation accuracy with the required speed of response. The mathematical provision of the commerically produced Dnepr computer does not allow one to solve this problem for the following reasons. First, the library of standard subprograms supplied with the machine allows one to conduct calculations with the error in the fourth to fifth places, whereas an accuracy two orders of magnitude higher was required for the solution of the stated problem. These subprograms are also rather cumbersome, which decreases the speed of response of the machine. Second, six-digit numbers can be inserted into the machine, although eight-digit numbers can be placed in its digital network. This fact automatically limits the potential accuracy of the calculations. Finally, the machine calculates only in a mode with a fixed decimal, which greatly hinders the work of the programmer and in a number of cases makes it almost impossible to program complicated algorithms having a large dynamic range of variation in the input parameters.

The IS-70 interpreting system and a library of standard subprograms for the calculation of elementary functions in three modes — with a fixed decimal, with a floating decimal, and in the mode of whole numbers* — were developed at the Ukravtomatika Trust in order to modern-ize the mathematical provision of the Dnepr computer. In particular, the new mathematical provision allows one to insert eight-digit numbers into the machine and to calculate elementary functions with an accuracy of six to seven places with sufficient compactness of the programs. Thus, the problem of providing the required calculation accuracy was solved. The required speed of operation was provided by the judicious combination of all three calculation modes in the programming of the algorithm.

An important positive feature of this system of automation of observations is the fact that by possessing considerable flexibility it is actually fully realized within the framework of the Dnepr computer, i.e., it does not require any external devices except for the timer and the feedback detectors. On the other hand, however, the construction of the system of auto-mation of observations on the basis of the Dnepr computer prevents one from analyzing the observational results in real time, except for the simplest operations of calculating the mean and dispersion, because of the relatively low operating speed of the machine (a cycle time of 4 μsec).

---

* Addition, subtraction, subtraction of absolute values, and multiplication are performed in the mode of whole numbers.

THE RTI-7.5/250 REFLECTING RADIO TELESCOPE WITH A FULLY STEERABLE PARABOLIC ANTENNA

P. D. Kalachev, V. P. Nazarov, A. A. Parshchikov.
and B. A. Rozanov

INTRODUCTION

The well-known achievements of observational radio astronomy in the last 10-15 years are to a considerable extent due to the equipping of radio astronomical observatories with the instrumental technology and primarily with the highly efficient radio telescopes of various systems and constructions.

The radio telescopes which presently exist differ from each other in dimensions and in structural schemes. This is explained both by the difference in the problems which must be solved in observational radio astronomy and by engineering and economical considerations.

However, statistics show that among the wide variety of structural schemes for radio telescopes the greatest popularity has been attained by reflecting radio telescopes with fully steerable parabolic antennas. This is explained by the advantages of telescopes of this type, namely: the needle-shaped directional diagram which permits the simultaneous two-dimensional resolution of the observed object; the stability of the directional diagram for different directions of the antenna in space, allowing one to conduct prolonged observations and to accumulate a signal; the capacity for simultaneous observation at different wavelengths through the use of matched exciters [1] suitable for polarization observations.

In addition, a fully steerable parabolic antenna is successfully used as an element of radio interferometers which permit the attainment of super-high resolutions [2].

It must be noted that the interest of radio astronomers in the millimeter wavelength range down to 1 mm has recently increased [3]. This explains the intensified work on designs for reflecting radio telescopes with fully steerable parabolic antennas for work in the short millimeter wavelength range being done in the USA (S. von Hörner) and Japan (Dr. Marimoto) and on the perfection of the existing radio telescopes in the millimeter range [3].

At the same time the creation of radio telescopes with fully steerable parabolic antennas, side by side with their advantages, is connected with great difficulties both in design and in fabrication. One of the main problems which must be solved in the creation of such radio telescopes is the problem of providing for the high focusing ability of the mirror system and maintaining it for any direction of the antenna in space. It is known that the greatest losses in the efficiency of parabolic antennas are caused by the deviation of the real reflecting surface of the main mirror from the ideal surface.

The effect of deviations of the real reflecting surface of the main parabolic mirror from the ideal surface on the size of the loss in efficiency of the parabolic antenna has been studied by many authors (see [4], for example). According to Ruze [5] (without allowance for the effect of the correlation radius) the size of the losses is determined by the equation

$$\Delta W = 1 - \exp\left[-(4\pi\bar{\varepsilon}/\lambda)^2\right],\tag{1}$$

where $\bar{\varepsilon}$ is the rms value of the error in the mirror surface and λ is the wavelength. The value $\bar{\varepsilon}$ includes the technological error of fabrication δ_t and the error (displacement) δ_d produced by deformations of the mirror:

$$\varepsilon = \sqrt{\overline{\delta_t^2} + \overline{\delta_d^2}}.\tag{2}$$

Reduction in the efficiency of the parabolic antennas of radio telescopes also results from other causes, namely: irregularity of the field in the aperture of the mirror, characterized by the aperture coefficient; reradiation; diffraction in the presence of a relatively large ratio $\lambda/D_{s.m}$, where λ is the wavelength and $D_{s.m}$ is the diameter of the secondary mirror; shading of the main mirror by structures of the exciter system; errors in the alignment of the elements of the mirror system (a discrepancy between the actual mutual arrangement of the main and secondary mirrors and the exciter proper and the calculated arrangement).

However, the main share of the losses in the efficiency of parabolic antennas is determined by the loss due to errors in the reflecting surface of the main mirror. According to statistical data for actual radio telescopes with fully steerable parabolic antennas the losses of efficiency are:

1) imperfection of the irradiation $\Delta W_{irr} \simeq 10\%$;

2) errors in alignment $\Delta W_{ali} \simeq 3\%$;

3) shadowing $\Delta W_{sha} \simeq 10\%$;

4) errors in the reflecting surface of the main mirror (imperfection in focusing)

$$\Delta W_{m.m} \simeq 10 \text{ to } 95\%.$$

And in fact, according to Eq. (1) for a ratio of the error in the surface to the wavelength $\bar{\varepsilon}/\lambda = 0.1$ the losses are $\Delta W \simeq 80\%$, for $\bar{\varepsilon}/\lambda = 0.05$ they are $\Delta W = 33\%$, and only for $\bar{\varepsilon}/\lambda = 0.03$ do the losses $\Delta W \simeq 13\%$ become acceptable. Thus, the principal reserve for an increase in the efficiency of parabolic antennas consists in decreasing the deviation of the reflecting surface of the mirror from the ideal surface, i.e., improving the focusing properties.

The current measuring methods and fabrication technology permit one to attain very small values of $\bar{\delta}_t$. As for elastic deformations, in previous constructions the $\bar{\delta}_d$ have been so large that parabolic antennas have had limitations in the capacity to work at short wavelengths for just this reason. For example, a radio telescope with an 86-m fully steerable parabolic mirror (Jodrel Bank, England) with a two-support fastening of the mirror operates [6] at the wavelength $\lambda_{min} = 21$ cm, and one with a 90-m parabolic mirror with rotation only in the plane of the meridian (Green Bank, USA) operates [7] at the wavelength $\lambda_{min} = 50$ cm.

It can be shown that in radio astronomy the main loads whose deformations distort the shape of the mirror are dead-weight loads. The point is that deformations from wind loads and especially from the effect of uneven heating by solar radiation can be either considerably reduced or completely eliminated. The decrease in wind loads and in the uneven heating by solar radiation is achieved by decreasing the time of the observations in accordance with the weather conditions: the observing time is limited to calm weather and overcast days. In addition, to decrease the effect of solar radiation the structures of the mirror system are painted in a white color (white titanium) [8] which reflects 90-95% of the thermal solar radiation. The complete elimination of the effect of external conditions is achieved by placing the radio telescope inside a dome which is transparent in the range of radio waves.

As for deformations from the dead weight, they cannot be eliminated in principle under the conditions at the earth's surface.

Deformations from the dead weight when the structural scheme is invariable are proportional to the square of the mirror diameter and essentially depend on the suspension (fastening) of the mirror on the supporting-rotating mechanism.

Generally speaking, one could reduce the distortion of the shape of the mirror produced by elastic deformations by increasing its rigidity. However, studies of the overall geometry of a mirror system subject to the effect of the dead weight have showed that even with an absolutely rigid mirror the problem is not completely solved since elastic deformations of the suspension of the exciter (secondary mirror) lead to defocusing of the mirror system during rotation of the antenna about the horizontal axis.

If one is to solve the overall problem of improving the focusing properties of a mirror system then one must, in addition to preserving the paraboloidal shape of the main mirror (homologous deformations), assure the elimination of defocusing of the mirror system or decrease it considerably. Defocusing of the mirror system arises during rotation of the antenna about the horizontal axis because of longitudinal (along the geometrical axis) and transverse displacements of elements of the exciter system. These displacements are due to deformations of the structures of the exciter system suspension. Longitudinal displacement of the exciter, causing broadening of the main lobe of the directional diagram and reducing its level, is especially troublesome. The mirror does not have to be absolutely rigid to prevent defocusing. The shape of the mirror must change in such a way that, while remaining a paraboloid of rotation, its focus will vary in accordance with the longitudinal displacement of the secondary mirror. Our studies of the elastic properties of a parabolic mirror showed that its deformations satisfying the conditions mentioned can be achieved by employing a multisupport mirror suspension with a radially symmetrical arrangement of the supports [9-11].

In connection with the fact that the elastic deformations of a mirror increase with an increase in its size in proportion to the square of the diameter the opinion has naturally been asserted that the problem of deformations is important for large mirrors, and this is valid to a certain extent. However, the problem of elastic deformations also exists for small mirrors. Optical telescopes are an example of this. It is known that the retention of the calculated (paraboloidal) shape of an optical mirror during rotation about the horizontal axis grows into a serious problem when the diameter is over 5 m. This is explained by the very rigid demands on the allowable deviations of the actual reflecting surface from the calculated surface. In optics these tolerances are tenths or even hundredths of a micron. As for the accuracy of fabrication, which is almost fully determined by the accuracy of the measurements (control), this problem was solved long ago in optics. Control of the reflecting surface of a parabolic mirror by the Foucault method [12] allows one to observe deviations in the surface of hundredths or even thousandths of a micron.

Evidently, control of the surface for parabolic mirrors of radio telescopes even in the range of short millimeter wavelengths will be rapidly perfected on the basis of the use of lasers. Therefore the problem of retaining the shape of parabolic radio telescope mirrors with shortening of the wavelengths used will again remain one of the most important problems.

Considering the above and having noted the trend toward the development of structural schemes for fully steerable radio telescope antennas for work in the range of short millimeter waves, in 1965 at the P. N. Lebedev Institute of Physics of the Academy of Sciences of the

USSR (FIAN) with the cooperation of the Bauman Moscow Technical College (MTC) we began the designing of a radio telescope with a fully steerable parabolic mirror 7.5 m in diameter intended for work in the millimeter wavelength range down to 1 mm.

1. STRUCTURAL SCHEME AND MAIN CONSTRUCTION

ELEMENTS OF THE RTI-7.5/250

The plan of the mirror system of the RTI-7.5/250 is based on a structural scheme with a multisupport suspension of the main mirror proposed and developed in the laboratory of radio astronomy (radio telescope sector) of the FIAN.

The structural scheme of the mirror system of this radio telecope includes three main subassemblies: the main parabolic mirror (force framework), an intermediate eight-support structure, and an eight-rod supporting pyramid with a ninth central rod [13]. Thus, the total number of supports for the main mirror is 17.

The main parabolic mirror with a diameter of 7.5 m and a focal distance of 3 m (F/D = 0.4) is made in the form of an all-welded framework of thin-walled steel tubes faced with aluminum sheeting* with a thickness $\delta = 3$ mm. The front side of the aluminum facing forms the reflecting surface of the mirror.

The framework of the mirror consists of a three-dimensional radially symmetrical rod system. It consists of 16 flat radial girders and the required number of chord girders. Knee braces are contained in the lower panels formed by belts of radial and chord girders. In the central part of the framework there is a central bushing containing a conical opening exactly machined on a lathe for fastening the axis of the rotating template. Nodes for fastening the mirror to the intermediate eight-support structure are welded to the lower belts of the eight-chord girders after every second radial girder. Thus, all the radial girders of the framework are arranged identically relative to the supporting nodes, which in turn are located radially symmetrically relative to the center of the mirror. These nodes form the outer stage of mirror supports.

The intermediate eight-support structure is also all-welded and made of steel construction tubes. It consists of four identical flat girders, each pair of which is located mutually parallel and intersects the other pair at right angles, forming a square in the middle part and the eight mirror supports at their end points. Knee braces are located in the upper and lower panels of the paired consoles of flat girders. The nodes for fastening the intermediate structure to the rotation sectors are welded to the four joints of the square formed by the intersecting girders.

On the consoles of two mutually parallel girders of the intermediate structure (symmetrical relative to the plane of rotation of the mirror in elevation and passing through the geometrical axis) brackets are fastened carrying the weight disks of the counterweight with which (through regulation of its arm and the number of disks) the retention of the plane arrangement of all eight supports of the outer stage is accomplished.†

---

* In a second improved variant the thickness of the facing was 6 mm.

† During rotation of the antenna each four supporting points of a single group always remain in one plane, but the two planes of supports of the two groups, generally speaking, do not remain in the same plane during rotation of the antenna in elevation. These weight disks, in accordance with the regulation of the moment which they produce, decrease or fully eliminate the departure of the two groups from a single plane.

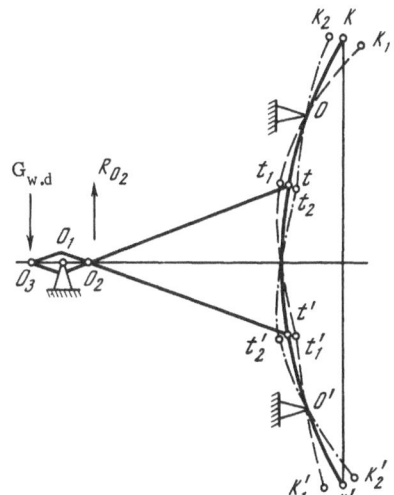

Fig. 1. Diagram of compensation for skew-
symmetrical deformations.

All eight supports of the intermediate structure have the same rigidity in the direction parallel to the geometrical axis of the mirror. The supports within a single group* have the same rigidity in the direction parallel to the aperture of the mirror and parallel to the plane of rotation of the mirror (in elevation). There are two such groups of supporting points and each group, generally speaking, has a different rigidity from the other.

The inner stage of supports is formed by an eight-rod supporting pyramid whose central (ninth) rod forms the central (17th) support of the mirror. The supporting points formed by the rod pyramid also have the same rigidity.

The subassembly of rotation sectors consists of two identical toothed sectors of the elevation drive (with a diameter of 4100 mm over the pitch circle), the horizontal axis with the supporting journals, and a system of tie rods connecting the two sectors (along the rims) and forming a kind of squirrel cage. To impart great rigidity against twisting to the subassembly of sectors, i.e., to prevent the mutual angular displacement of the sectors in their planes, knee braces are introduced into the planes of each pair of tie rods. The lower points of the sectors (on their axes of symmetry) are connected together by a reinforced tie rod — a tube on which the eccentric weight disks of the counterweight are located. The tie rod tube can rotate about its own axis.

A complete calculation of the main mirror for skew-symmetrical loading was not conducted. For symmetrical loading the calculated elastic deformations at the edges of the mirror from the dead weight are ~0.017 mm.

On the basis of an approximate calculation (without allowance for the compensating effects of the balance systems) the maximum deflections at the edges of the mirror and in the vertical cross section (antenna directed toward the horizon) are ~0.1 mm. However, the balance compensation systems provided in the structure decrease this deflection by about five times. The system of balance compensations of the deflections during skew-symmetrical loading of the

---

*We call the four supports formed by the end points of the consoles of two mutually parallel flat girders of the intermediate structure a group of supports.

main mirror consists of two units. The first unit is a device containing eccentric weight disks arranged along the horizontal tube joining the lower midpoints of the toothed sectors of the telescope elevation drive. The principle of operation of this unit consists in the following (Fig. 1): In the absence of a load at the point O_3 the supporting rod pyramid O_2-t-t' is free and does not act during skew-symmetrical loading on the mirror. In this case the mirror profile in the vertical cross section from the effect of the dead weight is deformed in the way shown by the dashed line in Fig. 1: The consoles are bent downward, the upper point t is displaced to the left to the position t_1 and the lower point to the right to the position t_1'; the upper end K of the console is displaced to the position K_1 and the lower end to the position K_1'. If the force R_{O_2} directed upward is applied to the apex O_2 of the pyramid then the point t is displaced in the opposite (to the previous) direction, i.e., to the right, to the position t_2 and the point t' is displaced to the left to the position t_2'; the upper end point K of the console is displaced to the left to the position K_2 and the lower end point K' of the console is displaced to the right to the position K_2' (Fig. 1, line marked with dots). Since by acting on the apex of the supporting rod pyramid one can cause the points K and K' as well as t and t' to be displaced in the directions opposite to those in which they are displaced in the absence of the pyramid, one can consequently select a force R_{O_2} for which these displacements are almost completely compensated for.

The force R_{O_2} arises from the component of the weight $G_{w.d}$ of the weight disks and is determined by the equation

$$R_0 = \frac{(O_3 O_1) \, G_{w.d}}{O_1 O_2} \cos \varphi,$$

where φ is the angle between the direction of the geometrical axis of the mirror and the horizon.

The second unit of the compensation system consists of a set of weight disks arranged on brackets on the back side of the mirror.

2. TECHNOLOGY OF MOUNTING AND EXACT ALIGNING
(FINAL ADJUSTMENT) OF FACING SHEETS FORMING
REFLECTING SURFACE OF MIRROR

a) Mounting of Facing Forming the Reflecting
Surface of the Mirror and Its Preliminary Placement
in the Designed Position

The facing sheets (aluminum 3 mm thick), preliminarily shaped with an accuracy of ~2-3 mm, are fastened to the framework by means of threaded regulating pins. The total number of such pins over the entire framework of the mirror is 4500. The pins are fastened to the framework (on the upper belts — the tubes of the radial and chord girders) with the help of clamps which are distributed on the belts at the required points and positions and are fastened with pinch bolts. The facing sheets are fastened to the pins with flat-headed screws (there are threaded openings for this in the pins). The maximum size of the aluminum facing sheets is 1200 × 1000 mm. The total weight of the facing is ~500 kg.

For the operation of the radio telescope at the wavelength $\lambda = 2$ mm it is necessary that the reflecting surface of the mirror be fabricated with a maximum error of no more than 0.2 mm. In the fabrication of a highly accurate parabolic surface the technological errors are determined mainly by the measurement errors. Therefore the method of the measurements had to be chosen first of all. Studies of the existing measuring methods showed that the most suitable is the method of measurements and control of the fabricated reflecting surface of a paraboloid of rotation using a rotating knife template. If one has a rotating knife template whose axis is rigidly fastened to the mirror framework then the accuracy of the mirror alignment is almost equal to the accuracy of fabrication of the template, with allowance for the errors produced by errors in the bearings on which the axis of rotation of the template is mounted.

b) The Rotating Knife Template. Construction and Method of Preparation

The rotating knife template consists of a rigid flat girder made of welded rolled steel. The flat girder is faced on one side with stainless steel sheeting 1.5 mm thick. The facing sheets are fastened to the girder with threaded regulating (in height) pins. Such fastening of the facing allows one to regulate it in a single plane with sufficient accuracy (the maximum deviation from the plane is about 0.5 mm).

The template is two-sided, i.e., its working edge is a "full" (two-sided) parabola arranged symmetrically relative to its apex which is in the center of the template. The girder of the template is divided and its two halves are joined to each other by means of a central body which in turn is fastened to the shaft of the axis of rotation. This shaft is milled in such a way that the plane of the central body, being a continuation of the plane of the template facing, coincides with the axis of rotation. Regulating screws are provided in the central body and in the axis shaft for the accurate placement of the template on the axis of rotation. Devices which eliminate the clearance in the bearings are provided in the construction of the bearing assemblies of the axis of rotation so that the template rotates without free play. The roller bearings were chosen from a group of increased accuracy. Measurements showed that the rotation of the template about its axis, which is also the geometrical axis of the mirror, takes place with high accuracy. The deviation of the end of the template from the horizontal plane (the axis of rotation is vertical) in which it must lie if errors are absent does not exceed ±0.03 mm during rotation through 360°.

The weight of the assembled template (girder, facing, central body, and shaft of axis of rotation) is ~1500 kg.

First of all the template was marked off according to the calculated coordinates of the generatrix of the parabola. For convenience in the marking a base line was created which served as the upper straight line of the facing and which was machined in accordance with a string and with two precision levels having an accuracy of ±0.02 mm. As a result of machining the working edge of the template along the marking an accuracy of ±0.15 mm was achieved, which is close to the accuracy of the templates made earlier during the creation of the RT-22 of the FIAN and the RT-22 of the Crimean Astrophysical Observatory [13, 14].

Since the attainable accuracy in the fabrication of the template was insufficient, a new method of measuring and finishing the working edge of the template using optical instruments was proposed and developed at the Bauman Moscow Technical College. First of all the base line was recalibrated using an OSK-2 optical bench (Fig. 2). The plane of the bench rails was set up using levels and an AKM-1000 autocollimator. The bench itself was placed along the plane of the template parallel to the upper straight edge. Two lines (reference lines) were made in the facing sheet by moving a stand with a cutting tool along the rails of the bench. The straight base line was newly machined along the upper line. The lower reference line served as a control. Now the straight base line was obtained with an error of no more than 0.05 mm.

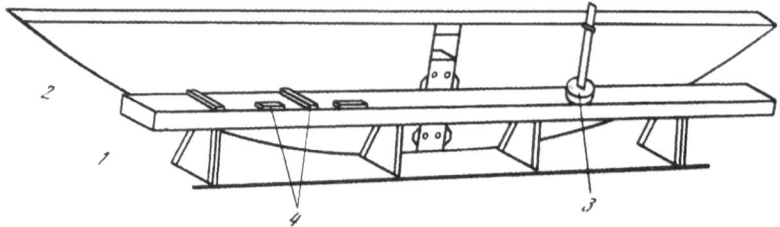

Fig. 2. Diagram of alignment of straight base line using an optical bench. 1) Optical bench; 2) template; 3) marking tool; 4) levels.

Control measurements conducted by geodesic means by colleagues of another institution confirmed the accuracy of ~0.05 mm obtained.

The further final adjustment of the template was performed by optical means on the basis of the use of the ability of a parabola to focus a parallel beam of rays traveling along the optical (geometrical) axis of the parabola onto one "point" (the focus of the parabola).

The method consisted in the following: a beam, sent from the light source 1 (Fig. 3) and reflected from the reflecting surface of the prism 2 placed at a 45° angle, is directed toward the controlled template edge which must fit a parabola. A plane mirror 3 abuts against the edge of the template, resting on it at the two end points of the mirror. Reflecting from the mirror 3, the beam is directed at the focus of the parabola (the screen 4). The error in the machined edge of the template is determined from the deviation of the beam from the focal point at which the screen 4 is mounted.

An LG-75 laser, which has good stability, a small beam width in the unimodal operation mode, low dispersion, and sufficiently intense radiation, was used as the light source.

The movement of the reflecting prism, which had to be strictly parallel to the straight edge of the template, was controlled with an AKM-1000 autocollimator on the face of the prism parallel to the geometrical axis of the template (see Fig. 3).

A test of the use of this method showed that the errors in machining the working edge of the template by this method are determined by the errors in the mutual arrangement of the

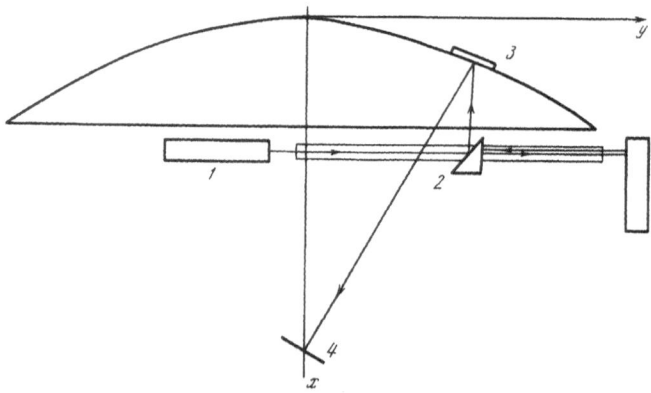

Fig. 3. Diagram of control of final adjustment of template by optical means.

elements of the measuring complex, i.e., by the positions of the AKM-1000, the reflecting prism, the screen (in the focal plane), and the mirror on the controlled edge of the template.

For an increase in the accuracy of the method introduced it is necessary to increase the accuracy in reading the deviation of the beam reflected from the edge of the template. This can be achieved by two methods: 1) by decreasing the size of the spot of the reflected beam on the focal screen; 2) by increasing the scale of the reading system.

In the first case attention must be turned to the mode of operation of the laser and the quality of the surfaces of the optical elements (the prism and the bridge mirror). The use of a spherical mirror on the reflecting bridge focusing the laser radiation onto the screen and the use of a diaphragm are possible. Even with the smallest diameter of the focal spot, however, it is difficult to resolve deviations of the beam of less than 0.3-0.5 mm.

The second course is more reasonable. One of the variants of its realization is illustrated in Fig. 4, from which it is seen that

$$\Delta S = \sqrt{(f-x)^2 + y^2} \tan 2\varphi, \qquad \sin \varphi = \frac{\Delta l}{R},$$

$$\Delta S = \sqrt{(f-x)^2 + y^2} \, 2 \frac{\Delta l}{R}.$$

Hence

$$\Delta l = \left(f + \frac{y^2}{4f} \right) 2 \frac{\delta}{R}$$

or

$$\Delta S = \left(f + \frac{y^2}{4f} \right) 4 \frac{\delta}{dR}; \quad \ldots \tag{3}$$

With R = 50 mm

$$\Delta S = \pm 1 \text{ mm}, \qquad \delta_{\max} = \delta_{y=0} = \frac{R d \Delta S}{4 f^2} = \frac{50 \cdot 50 \cdot 1}{4 \, (3.25 \cdot 10^3)^2} = 6 \cdot 10^{-6} \text{ mm}.$$

The value obtained for the measurable error is considerably smaller than the allowable error in the fabrication of the working edge of the template.

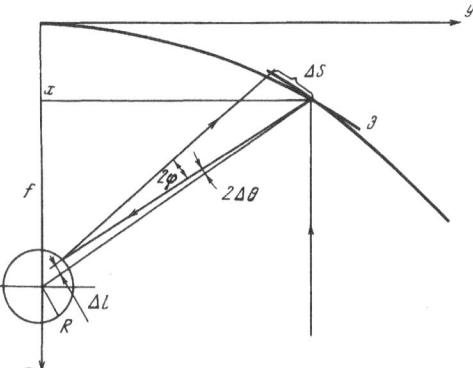

Fig. 4. Diagram of the increase in the accuracy of the optical method of control through an increase in the scale of the reading system.

The accuracy of fabrication of the template finished by the method described was ~0.017 mm (rms value of the error).

c) Final Alignment (Adjustment) of Reflecting
Surface of Mirror

The final alignment-adjustment of the reflecting facing of the mirror is carried out with the help of the rotating knife template which is placed on the mirror with the latter directed toward the zenith. The axis of the template is fastened in the conical opening of the central bushing of the framework. The deflection of the mirror at the center under the load of the template's weight ($G_{tem} \simeq 1500$ kg) was measured with a gauge (during the installation of the template) and was equal to ~0.3 mm. To restore this deflection the central part of the mirror was supported with a jack which raised it by ~0.3 mm. The calculated deflection of the end of the template by ~0.08 mm agrees with the deflection measured experimentally.

Fig. 5. Canvas tent over the mirror system.

The sequence of operations during the measurement of the mirror surface using the template is as follows: before the placement (mounting) of the facing sheets the regulating pins are set exactly with respect to the template, with a clearance gauge 3 mm thick (equal to the thickness of the facing) being placed between the upper end of the pin and the measuring edge of the template; the clearance between the working edge of the template and the front side of a facing sheet is measured at a sufficiently large number of points of the sheet by the rotation of the template within the limits of one detachable sheet; the sheet is removed from the framework and its additional correcting is carried out on a matrix having the shape of the paraboloid at the given location; the sheet is fastened to the pins with flat-headed screws; the gap between the facing sheet and the edge of the template is measured and the difference in the values obtained is removed by the method of tin-coppersmithing procedures (using wooden and rubber hammers); the final adjustment of the surface is done by scraping.

A test showed that the adjustment of the surface with respect to the template by scraping with an accuracy of ±0.1 mm is a very laborious operation. Therefore in the first variant of the mirror because of a lack of time (it was late fall) it was decided to reduce the accuracy of the surface adjustment by allowing an error of ±0.2 mm.

For convenience in performing the work on the adjustment of the reflecting surface the mirror system was covered with canvas walls and roof which shielded the mirror from the effect of the direct solar rays and partially from the wind. Moreover, this cover made it possible to work during a drizzling rain when the temperature conditions were judged the most favorable (Figs. 5 and 6).

Fig. 6. Working conditions under the tent.

d) Control of Fabricated and Adjusted Reflecting
Surface of Mirror and Determination of Total Error

The total technological error in the fabrication of the surface can be represented in the form

$$\Delta_r = \sqrt{\Delta_{meas}^2 + \Delta_{tem}^2 + \Delta_{ax\ tem}^2},$$
(4)

where Δ_{meas} is the error in measurement by the template, Δ_{tem} is the error in the fabrication of the template, and $\Delta_{ax\ tem}$ is the error in the rotation of the template on the axis (the inaccuracy of the bearings).

Although the measurement of the mirror surface by the template is considerably simpler and more reliable than measurement by the geodesic method, this operation is nevertheless very laborious. Therefore we proposed and developed a new original means of group measurement of the reflecting surface relative to the working edge of the template using a group of four pickups fastened to the structure of the template (Fig. 7). These linear displacement pickups (LDP) consist of differential induction pickups developed especially for such measurements. The signals from the pickpus are sent to an amplifier and then to a recorder. A series-output eight-channel 8ANN4-7M strain station and an N700 loop oscillograph with recording on photographic paper were used for the amplification and recording of the pickup signals. Before the measurement the channels of the measuring system as a whole were calibrated with a special calibrating device.

Fig. 7. Method of group measurement of reflecting surface relative to edge of template by a group of four pickups.

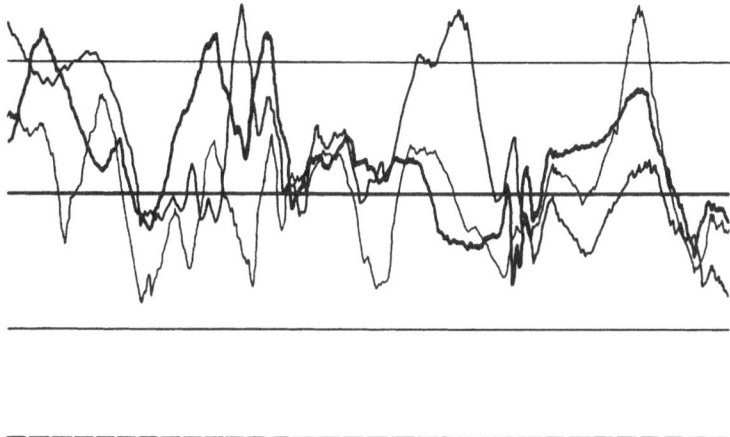

Fig. 8. Oscillogram of readings of control pickups.

The effect of instability in the amplification of the strain station, caused by the conditions of work with fluctuation in the temperature and supply voltages, is eliminated by the periodic feeding to the inputs of the strain station of calibrated signals which are recorded in the form of reference marks along with the LDP signal (Fig. 8).

During the rotation of the template about the geometrical axis the surface was measured on four concentric circles at once in each revolution. After each revolution of the template the group of four pickups (the pickup block) was shifted along the edge of the template to a new position and fastened to the structure.

For a control on the surface measurements obtained along all the circles (the total number of such circles over the entire mirror was 35) measurements were made at the points of intersection of these circles with one of the radial directions (along the edge of the stationary template). These measurements showed that the deviations (in the range of 0.02-0.04 mm) of the surface from the template edge do not exceed the deviations obtained in the circle measurements.

It must be noted that the mounting of the pickups relative to the measuring edge of the template must be performed with particular care (in both cases of measurement: both in the movement of the template in a circle and in radial displacements of the group of pickups along the template) since this determines not only the quality (reliability) of the coordination of the measurements in both directions (along the circle and along the radius) but also the overall reliability of the results obtained. Therefore the mounting of the groups of pickups was accomplished with a special adapter consisting of a milled try square, one of whose flanges (the lower horizontal flange) was pressed against the working edge of the template (on the face) while the other adjoined the plane of the edge. The measuring probe of a pickup rested against the horizontal (lower) flange of the try square and there its reading was set at zero. When conducting the measurements the adjusting try square was retracted and the probe touched the reflecting surface of the mirror, i.e., the front side of the facing. During the rotation of the template about its axis the probe duplicated all the irregularities of the facing, which was recorded by the oscillograph on photographic paper (Fig. 8).

e) Automatic Control of Mirror Surface

The preliminary data on the analysis of the results of measurements of the mirror surface using the strain stations consumed a considerable amount of time.

Assuming that the control work connected with the expansion of studies of the behavior of the mirror in the course of operation would increase, it was decided to automate the measurement process in the recording of the results. This was facilitated by the principle of operation of the grouped linear displacement pickups described above, which produced signals in the form of an electrical voltage. A system using the Dnepr-1 digital computer was proposed and developed. The voltage signals from the linear displacement pickups, proportional to the deviations of the mirror surface being measured, are fed to the input of the computer using a coupling device. The signal $X^j(t)$ of the j-th LDP enters the analog input of the machine where, in accordance with the program developed, it is made discrete in time and divided into 256 levels, being converted into the random value kX_i^j (k determines the specific sheet of the facing and j determines the number of the LDP).

Then the sequence of discrete values kX_i^j is processed in real time to obtain the mean value, dispersion, and probability density of the deviations both for the individual sheets of the facing and for an individual (j-th) circular cross section of the mirror as a whole.

The characteristics analyzed were calculated from the following equations:

$$a^j = \frac{1}{n}\sum_{k=1}^{n} a_k^j, \qquad a_k^j = \frac{1}{aN}\sum_{i=1}^{N} kX_{ic}^j, \tag{5}$$

$$\sigma_k^j = \sqrt{\frac{1}{a^2 N}\sum_{i=1}^{N}(kX_{ic}^j)^2 - (a^j)^2}, \tag{6}$$

$$\sigma^j = \sqrt{\frac{1}{n}\sum_{k=1}^{n}\frac{1}{a^2 N}\sum_{i=1}^{N}(kX_{ic}^j)^2 - (a^j)^2}, \tag{7}$$

$$W_k^j(X_l) = \frac{m_{lk}^j}{N}, \qquad W^j(X_l) = \frac{1}{n}\sum_{k=1}^{n} W_k^j(X_l),$$

where $kX_{ic}^j = kX_i^j - X_0^j$; $X_0^j = X_k^j - \alpha\Delta_k^j$; $\Delta_k^j = \delta_k^j - \delta_0$; $\alpha = X_{cal}/\Delta_{cal}$ is the calibration coefficient, m_{lk}^j is the number of readings with the condition that $X_{l-1} < kX_{ic}^j \le X_l$ (l = 1, 2, ..., 17); $X_l = (l-9)\,\Delta X$, $\Delta X = 2^4$ is the interval of division in the calculation of the histogram of probability density of the deviations; δ_k^j is the deviation of the mirror surface from the working edge of the template at the control point, measured with high accuracy; δ_0 is the nominal deviation of the mirror surface from the template (the average of the technological field of deviation, in our case 1.8 mm); X_k^j is the LDP signal at the control point; n = 16 is the number of sheets in a level; N = 150 is the number of readings per sheet in one cross section.

The frequency of making the signal discrete is assigned by a special angle detector, which puts out the signal determining the discreteness after rotation of the template through a fixed angle θ (θ = 10' when N = 150).

The sequence of operations of the control method is as follows. The initial data are Δ_k^j and Δ_{cal}. The X_{cal} which corresponds to Δ_{cal} is given before the start of the control session; the machine calculates α on the basis of the value of X_{cal}. Then the template is set so that the LDP stand at the control points and X_k^j is entered into the machine, after which X_0^j is calculated (this operation takes place almost instantly) and the control of the surface is performed.

The entire process of the measurements is carried out during the rotation of the template through 360°; after the rotation of the template through an angle corresponding to one sheet of the facing the characteristics considered above are printed out. The characteristics corresponding to the specific cross sections of the mirror are printed out when the template has completed a full revolution.

3. THERMOELASTIC DEFORMATIONS OF MAIN MIRROR

Although in radio astronomy one can within certain limits confine the observation time to favorable weather conditions, the desirability of increasing the time of use of the radio telescope (the "efficiency") forces one to seek means of overcoming the harmful effect of the weather conditions, particularly, as mentioned above, to decrease the effect of thermal solar radiation in distorting the geometry of the mirror system and first of all the shape of the main mirror. Therefore it becomes necessary to at least estimate the thermoelastic deformations of the main mirror, which can be considerable [15]. For this one must determine the degree and nonuniformity of the heating of structural elements of the main mirror and its suspension, i.e., find the temperature field in the bearing framework of the mirror.

A method of monitoring the temperature distribution over the structure of the mirror system was proposed and developed at the MTC for this purpose. The basis of the method is the use of a measuring bridge circuit. A "search" thermistor of the MMT-4 type is incorporated into one of the arms of the bridge circuit through a step commutator. The readings of the pickups (thermistors) are recorded on the tape of an ÉPP-09 recorder. The balance of the bridge is regulated for the variation of the range of temperatures recorded.

The temperature was measured at 40 points located at the most characteristic positions of the structure of the mirror system, at which the 40 thermistors were mounted, namely: on the outer suspensions of the 16 radial girders, at the nodes for fastening the intermediate supporting structure to the sectors and to the mirror framework, at the fastening points of the upper ends of the eight-rod supporting pyramid, on the central bushing, and on the facing. The temperature of the surrounding air was monitored simultaneously. The length of a cycle of interrogation of all 40 points was 12 min.

An analysis of the operation of the measuring system showed that the error in recording the temperature did not exceed ±0.2°C.

On the basis of the results of the measurements the following was established:

1) The greatest temperature differential was observed during solar heating in periods of rapid changes in the temperature of the surrounding air;

2) Direct solar heating produces the main effect, and the effect of a moderate wind and rare clouds is unimportant;

3) The temperature of the structural elements located in the shade is close to the temperature of the surrounding air. This indicates the low heat transfer from the more heated to the less heated elements;

4) The facing was subjected to the greatest heating: Its temperature exceeded that of the surrounding air and the shaded elements of the structure by 15-20°.

On the basis of these data one can make the following recommendations for decreasing the temperature differential in elements of the mirror structure: Encircle the mirror along the border with a screen which shades its body from the direct solar rays at angles of ±50° from the direction toward the sun.

Fig. 9. General view of assembled RTI-7.5/250 radio telescope.

The small heat transfer from the facing sheets to the framework is a positive feature of the structure of the fastening of the sheets to the framework. This feature was noted earlier in [15] with regard to the RT-22 of the FIAN [13].

In addition, attention must be turned to the marked thermoelastic deformation of the aluminum facing sheets from the large temperature differential between the facing and the framework. In order to decrease the effect of this factor on the reflecting surface of the mirror it is evidently necessary to decrease the size of the facing sheets.

CONCLUSION

As a result of the great research, design-construction, and experimental work of the colleagues of the FIAN and the MTC, a radio telescope has been created which is unique in resolving power and which has a relative surface error of $\varepsilon/D = 0.15/7500 = 2 \cdot 10^{-5}$ in the first variant with a guaranteed reserve of increase in accuracy to $\varepsilon/D = 0.05/7500 = 7 \cdot 10^{-6}$

and in the second variant having a facing forming the reflecting surface of the mirror with a thickness $\delta = 6$ mm.

On the basis of the preliminary tests the efficiency of use of the parabolic mirror at the wavelength $\lambda = 2$ mm is 40%.

A photograph of the assembled RTI-7.5/250 radio telescope is presented in Fig. 9.

LITERATURE CITED

1. I. V. Vavilova, G. K. Galimov, P. D. Kalachev, A. M. Karachun, A. D. Kuz'min, B. Ya. Losovskii, and A. E. Salomonovich, Vop. Radioélektron., Ser. Obshchetekhn., No. 1, 13 (1964).
2. J. J. Broderick, V. V. Vitkevich, D. L. Dzhonski, V. A. Efanov, K. I. Kellerman, B. G. Clark, L. R. Kogan, V. I. Kostenko, I. G. Moiseev, D. Payne, and B. Hansen, Inst. Kos. Issled., Preprint Pr-20, (1970).
3. J. R. Cogdell, J. J. G. McCue, P. D. Kalachev, A. E. Salomonovich, I. G. Moiseev, J. M. Stacey, E. E. Epstein, E. E. Atschuler, G. Feix, J. M. B. Day, H. Hvatum, W. J. Welch, and F. T. Barath, IEEE Trans. Antennas Propag., AP-18, No. 4 (1970).
4. S. Ya. Shifrin, Problems of the Statistical Theory of Antennas [in Russian], Sovetskoe Radio, Moscow (1970).
5. J. Ruze, Proc. IEEE, 514 (1966).
6. A. G. Lyne, F. G. Smith, and D. A. Graham, Rep. Mon. Not. Roy. Astron. Soc., 153, 337 (1971).
7. G. Sinigaglia, Alta Frequenzy, 34, 581 (1965).
8. B. H. von Grahe, Die Sterne, 2, 65 (1972).
9. P. D. Kalachev, Tr. FIAN, 47, 77 (1969).
10. P. D. Kalachev, Preprint FIAN, No. 171 (1968).
11. P. D. Kalachev and M. V. Konyukov, Preprint FIAN, No. 149 (1971).
12. D. D. Maksutov, Astronomical Optics [in Russian], Gostekhizdat, Moscow-Leningrad (1946).
13. P. D. Kalachev and A. E. Salomonovich, Tr. FIAN, 17, 18 (1962).
14. V. N. Ivanov, I. G. Moiseev, and Yu. G. Monin, Izv. Krim. Astrophys. Observ., 38, 14 (1967).
15. Yu. L. Shakhbazyan, Izv. Glav. Astron. Observ., Part 3, No. 17, 2 (1964).